CW00819056

THE "WANDERER." *(From a picture by Admiral Beechey.)*

# THE VOYAGE

OF

# THE "WANDERER"

*FROM THE JOURNALS AND LETTERS*

OF

C. AND S. LAMBERT.

EDITED BY GERALD YOUNG.

ILLUSTRATED BY R. T. PRITCHETT, AND OTHERS.

London
MACMILLAN AND CO.
1883

LONDON: R. CLAY, SONS, AND TAYLOR, BREAD STREET HILL.

TO

THE MANY FRIENDS WE MADE

DURING OUR VOYAGE IN THE "WANDERER,"

WHOSE FACES IN THIS WORLD WE MAY NEVER SEE AGAIN,

BUT WHOSE

KINDLY HOSPITALITY WE CAN NEVER FORGET,

These Pages are Dedicated.

# PREFACE.

The following pages are compiled from the Diary kept by Mr. Lambert during the voyage of the *Wanderer*, and the Letters written by Mrs. Lambert to friends at home, but, for convenience' sake, the whole narrative has been put into the mouth of Mr. Lambert. The editor desires to acknowledge with gratitude the assistance he has received from Mr. Wetherall's letters, which have also been placed at his disposal.

Book-making has not been attempted, nor have guide books or other works of reference been consulted. The journal is a perfectly simple one, pretending to no literary merit or to any competition with successful publications of a somewhat similar character. It is published in the first place as a record of two very happy years spent afloat, which cannot fail to interest the relations and friends of the "Wanderers;" and in the second is intended to mark their grateful recollections of the kindness and hospitality shown to them by various nationalities and classes in all parts of the world.

GERALD YOUNG.

# CONTENTS.

# LIST OF ILLUSTRATIONS.

## COLOURED PLATES,

ENGRAVED AND PRINTED BY R. CLAY, SONS, AND TAYLOR.



## WOOD ENGRAVINGS,

ENGRAVED BY J. D COOPER.

## MAP.

# VOYAGE OF THE "WANDERER," R.Y.S.

Sailed from Cowes, Aug. 5th, 1880. Returned, July 19th, 1882.

Duration of voyage, 1 year, 348 days.
Of which 1 „ 68 „ in harbours.
And 280 „ at sea.
Distance traversed, 44,890 geographical miles.
Of which 31,015 „ „ under steam, or steam and sail.
„ 13,875 „ „ „ canvas only.
Average speed per diem, 160·32 geographical miles.
„ „ hour, 6·68 „

Total coal consumed during the voyage, including many days and nights, over fifty, under banked fires, in dangerous open roadsteads, and during the use of sails only, propelled by winds of doubtful duration, and including also coal used by two galley-fires, and consumed in calms, Tons 1,265 10 cwt.
Average consumption per diem on total sea days „ 4·52
„ „ „ „ 189½ steaming „ 6·77
„ „ „ mile traversed 0·62 „

*Persons on board when leaving Cowes.*

1. Charles J. Lambert, Owner and Master, without certificate.
2. Mrs. Lambert.
3. Helen Mark Lambert.
4. Beatrice Kate Lambert.
5. George Maximiano Lambert.
6. William Stanley Lambert.
7. Rev. H. E. Wetherall.
8. Robt. T. Pritchett, Artist.
9. Miss Julia Power, Governess.

b

10. Agnes McAllan, Ladies' Maid.
11. Elizabeth Cordell, Nurse.
12. John Harris, Valet.
13. John Dadge, Footman.

*Officers and Crew.*

1. A. Gordon, Sailing Master and Navigator.
2. Ed. F. Tyacke, Chief Officer.
3. John Williamson, 2nd ,,
4. John Boniface, Chief Engineer.
5. Henry Johnson, 2nd ,,
6. Arthur Richardson, 3rd ,,
7. J. Grey, Surgeon.
8. Henry Barnes, Carpenter.

1. Wm. White, Chief Boatswain's Mate
2. George Flowers, 2nd ,, ,,
3. Robt. Hooper, Quarter Master.
4. Rich. Powell ,, ,,
5. Sam. Pridham ,, ,,
6. Wm. Blight ,, ,,
7. John Cobbledick, A.B.
8. John Saunders ,,
9. George French ,,
10. Philip Angel ,,
11. Wm. Riddols ,,
12. John Foster ,, and Gunner.
13. Rich. Honeychurch ,,
14. Wm. Shepherd ,,
15. Otho Clarke ,,
16. Philip Towell ,,
17. Sam. Fletcher ,,
18. Reuben Ruby ,,
19. Robt. White. ,,
20. John Eckley ,,
21. Wm. Baskerville ,,
22. Edwd. Davy ,,
23. John Doidge ,,
24. James Bowley ,,
25. Wm. Kellan ,,
26. Nicholas Dudley ,,
27. John Gates, Boy afterwards O. S.

28. Thomas Furnace, Boy.
29. Charles Poulter „
30. Wm. Gould „
31. Peter Hendricksen, Leading Stoker and E. R. Storekeeper.
32. John Nicholas, Stoker.
33. Richd. Emmett „
34. Wm. Taylor „
35. Wm. Ruffin „ and Ship's Butcher.

*Stewards.*

1. Tho. Butler, Chief Steward.
2. Louis Busatil, Chief Cook.
3. Salvo Magrè, Assistant Cook.
4. Wm. Hollis, Baker and Pastrycook.
5. Roderick Morrison, Pantryman.
6. John Rust, Bedroom Steward.
7. Wm. Yeo, Officers' Servant and 2nd Steward.

Sixty-three persons, all told.

*Accidents on the Voyage.*

Upon leaving Tonga Tabu, Jno. Saunders, A.B., whose turn it was to go into the chain-locker to stow the cable while the ship was getting under way, either in a fit or from faintness, must have fallen forward, and, without a sound having been heard from him, was crushed and killed by the incoming cable, at least ten fathoms of which must have fallen over him before his position was discovered, when all hope of recovery was gone. The poor fellow was crushed fearfully, and must have died instantly.

On the 11th March, 1882, between Colombo and Suez, while running before the trade wind lazily, John Gates, boy, fell off the foretop and went overboard, but was picked up within six minutes by the life-cutter, with no other injury than a good fright and a good wetting, which would not trouble him much in this temperate and glorious climate.

*Changes in Personnel during the Voyage.*

<table>
<tr><td>Left the ship at Sandy Point, Straits of Magellan—<br>William Kellan, A.B.<br>Nicholas Dudley „<br>Tho. Butler, Chief Steward, at Coquimbo.<br>Arthur Richardson, 3rd Engineer, at Papeete.<br>Sam. Fletcher, A.B., at Yokohama.<br>Robt. Hooper, Qr.-master, at Hongkong.<br>William Taylor, Stoker, invalided at Colombo.<br>Louis Busatil, landed at Colombo.</td><td>Saunders, as before stated, died at Tonga Tabu, and with these changes excepted, all who left Cowes came home with us.</td></tr>
</table>

*Shipped at Coquimbo from England.*

Our son, R. S. Lambert, joined us here from England and came home in the ship.

| | |
|---|---|
| Wm. Leary, A.B.<br>Alex. Keswick "<br>Wm. Earle "<br>John Weldon "<br>John Williams " | There were six A.B.'s ordered out, but one poor fellow died on the way out. |

| | |
|---|---|
| Rbt. Field, Officer's Servant, shipped at Coquimbo.<br>Octave Desneux, Chief Cook, shipped at Honolulu.<br>Wm. Hamilton, 3rd Engineer " "<br>George Phillips, stoker, shipped at Colombo. | At Hongkong Wm. Knight and — Coulter joined the ship. |

9 left the ship on the voyage and 12 joined us, so that we returned to Cowes 66 strong.

# VOYAGE OF THE "WANDERER"

## CHAPTER I.

*FROM COWES—TO VIGO AND LISBON.*

THE *Wanderer* commenced her voyage from Cowes on the 5th August, 1880. The previous week had been pleasantly spent in receiving visitors and saying good-bye to kind friends and relations who had come to wish us good speed. Unmooring at 2. P.M., we steamed away to the west, accompanied by the *Lightning* steam tug which was engaged to take our visitors off. "Farewell and good wishes" were signalled from R.Y.S. Castle, which we duly acknowledged; and a little further on we passed the *Sunbeam* under canvas, following the racers, and got three hearty cheers from all on board her, and an "Adieu" by signal. At 5.30 P.M. we stopped off Yarmouth, to let the *Lightning* come alongside and take our dear visitors from us. This was the last wrench, and the moment that many on board, however bravely they might be keeping up, must have dreaded. A sad good-bye to each, a hearty God speed you, a cheer, the waving of hats and handkerchiefs, and the two vessels parted, the *Lightning* steaming back to Cowes, and the *Wanderer* at full speed wending her way to the westward.

The morning of August 6th found us at 4.15 A.M. with the Start light abeam, and at 7 A.M. we lost sight of England, the sea rising with a strengthening breeze which very soon had its effect upon our party—masters, servants, stewards, and even some of the men and boys succumbing to the demon of sea-sickness. At 10 A.M. we had run 160 miles from the Needles; at midnight the wind had veered from N.W. to S.W. in a fresh gale with a high sea, making the ship roll heavily, and obliging us to take in the fore staysail, and increasing the miseries and discomfort of those below. At noon on the 7th there was a strong gale blowing with heavy beam sea, ship taking in quantities of water fore and aft; at 6 P.M. however both wind and sea began to moderate, and at midnight, though the sea still ran high, only a breeze was blowing; and on the 8th at 8.30 A.M. land was sighted on the port bow: this news, with the bright sunshine and now fast falling sea, with the busy sounds of deck washing in exchange for that of the sea breaking on board, cheered up the invalids, and brought from the greater number a ravenous call for food again. Luckily Louis the cook was well and chirpy, and had been so throughout, although his mates were down, he had smoked his pipe and taken a holiday, no demands having been made on him either fore or aft. Wetherall had made several plucky attempts to get upon his legs during the gale, but had failed dismally, and even the cat, the sailors said, had been sick. The brilliant beauty of the day as it went on soon made all forget their sufferings. At one o'clock Coruna with its Tower of Hercules was abeam, and with a bright warm sun and clear blue sky overhead, with a light summer breeze from the N.W. and N., we continued running sufficiently close along the land to see the houses and very hedgerows. The sea was smooth, its crystals sparkling in the sunlight, and as we walked our clean dry deck again we felt well rewarded for the past two days of storm and discomfort, and that the darkness and tempest of the Bay of Biscay had only made us relish all the more the calm tranquillity of the shores of sunny Spain.

At 10 P.M. we sighted Cape Corrobedo, and at daylight on the morning of the 9th entered the beautiful bay of 9th Aug. 1882 Vigo, anchoring at 6 A.M. close under the town in nine fathoms of water. The day turned out piping hot, and we made good use of it by loosing and drying all our wet sails, and having all the bedding and damp clothes up. In the afternoon we manned the galley and the life-cutter, and getting masts and sails up, set off, with a pleasant breeze, I in the galley with our four children, and Captain Gordon and Wetherall in the life-cutter, to test the sailing qualities of the respective boats, as well as to take a look at the shores of the bay. The galley proved the fastest boat, as she should be from her lines and build; and all returned on board with well sharpened appetites. The harbour is certainly very fine, large enough to hold not only a fleet but many fleets, and may well be the favourite resort of our Channel and Reserve Squadrons to rest in between periods of steam tactics and other naval evolutions and manœuvres so much affected and so necessary in these days of ironsides and monster guns; although it seems as if it must be well nigh impossible to make ships efficient in all essentials—fast as a steamer, weatherly as a sailing ship, and strong both in offence and defence. The bay and town have been so well described by others, notably by Lady Brassey, that it is almost impossible to say anything new on the subject. The town on the southern side of the bay is very picturesquely situated on the side of a sugar-loaf hill capped with an old citadel—a useless stronghold in these days, however capable it may have been of offering a stubborn resistance at the beginning of this century. The formation of the land and scenery round the bay is not unlike that of many of the lochs on the Western Highlands of Scotland, although the hills are more conical and less rocky, the line above the growth of vegetation more parched, and lacking the purple heather. Here are waving fields of Indian corn, dark olive groves and vines, in the place of the birch, scanty oats, or root crops of the Highlands. The ground is in some parts thickly studded with tents and little churches, the first well stocked with ragged urchins and beggars, reminding one of the western portions of Cork and Kerry, although children, fishermen, and cottiers seem far better off than their Irish prototypes; and so they should be, for although their soil is not a rich one, it isn't bog—and as to climate, why all seems to go against the poor Irish and to make their lot a hopeless one. Still they cling to their huts and their mire, as do the Gallegos to their over-populated but far brighter homes, whose youth overrun Spain and her colonies in the same manner that the Irish have spread over England, the British Colonies, and the United States. Justly may the Gallegos be called the Irish of Spain, resembling each other as they do in religious bigotry and filthy habits.

On the 10th, before the sun gained much power, we paid a visit to the market-place, and were amused at first by the incessant din of the vendors and purchasers, each one striving to screech louder than the other; the clamour could be heard from dawn to dewy eve from the deck of our ship. We made up our minds that the good Gallegos must divide themselves into watches to keep up such a steady noise. We were glad to get out of it by 8 A.M., and to the quiet of our floating home. In the cool of the evening we went away in the cutter and tried to do a little fishing, but were not very successful, the total bag being only three mullet. Next morning, the 11th, we paid a visit to the cemetery, and 11th Aug. returning to the ship, at 10.35 weighed anchor and proceeded on our way to Lisbon. At the mouth of the harbour, we suddenly entered a dense fog, had to stop the ship, blow the fog-horn, and try the "steam siren," which gave forth the most awful grunts and squeals, and although we could hear other horns sounding, nothing was to be seen but the twinkling water close around us. We waited about half an hour, during which time what looked at first like a big ship appeared close to us, and passed round us; it was only a fishing boat, but the men with their slouched hats, dark bearded faces, and bright-coloured shirts seen through the fog had a most curious effect, and reminded one of a pirate ship as seen at the Opera Comique at home. At 11.35 A.M. we were able to go on again, getting calm and beautiful weather all the way down to Lisbon, and passing the little Moorish castle of Belem, cast anchor in the man-of-war ground at 1.10 P.M. on the 12th. In the afternoon we 12th Aug. all went ashore and secured the services of T. A. Franco, an excellent interpreter and courier. He took us for a walk into

Gold Street aud "Rolling Motion Square" the rough mosaic floor of which makes the flat surface present a peculiar wavy look. Thence to the fruit market, and back to the galley and on board in time for dinner.

h Aug. Next morning we were all bent on inspecting the sights of Lisbon, and left the ship after breakfast, finding Franco waiting for us with two carriages. We went to the Cathedral, then to the Church of San Vicente with its adjoining mausoleum, containing tombs of the kings and prnces of the House of Bragança from John IV. down to the present day, then on to the ruins of the old Cathedral, destroyed by the terrible earthquake. Many of the marble columns and arches have been replaced, and one can get a very good idea of what the beauty of the structure must have been. The site is now used as an archæological museum, the contents of which were fully and kindly described to us by the Chevalier de Silva. From here we went to the Chapel of San Roque, which contains three magnificent mosaics from paintings by Rubens; these alone make Lisbon well worth visiting. It was now luncheon time, and we were not sorry to sit down and rest a little, and discuss what we had seen. The town of Lisbon seems to be built upon a number of hills, three of which face the anchorage and are covered with churches and houses, skirted at the water-side by public buildings, and the quay well lined with ships of all sizes; in the background the sunlit river, on which craft, with lofty lateen yards and sails skim swiftly to and fro, forms a bright and animated picture.

The streets in the lower town are wide and well pitched; on the hill-sides, however, they become very steep and narrow, but the balconies full of brilliant flowers, and the gay-coloured garments hanging on them give a very picturesque appearance. If you want to see the sights of Lisbon you must go up and down streets of seemingly impossible gradients, for its churches, gardens, and old forts are perched on different hills, so you have to go up one steep side and down the other to reach the various points of interest. The more pretentious buildings are built of a good limestone or coarse marble; others are faced with a blue and white glazed tile from top to bottom. The roof is usually of red tiles. The poorer classes seem well housed, living in flats of four or five stories high.

After lunch we made a steep ascent to reach the Paseo and garden, or rather shrubbery, bright in varied colours, and quaintly interspersed with sun-flowers of a size and brilliancy of colour none of us had ever seen before. In the centre was a fountain surrounded by water-carriers of both sexes, dressed in bright-coloured garments, and carrying on a ceaseless chatter. From here an extensive view is obtained of such parts of Lisbon as are not hidden by the numberless hills; a fine view of the wide and winding Tagus is also obtained from this Paseo or promenade. We descended the hill, passing through the fine Plaza da Principa Imperial, and up again to the so-called Botanical Gardens, faced on one side by a large square building which contains the Polytechnic School, in which cadets for the army, navy, and military engineers are educated; on from here to the waterworks, situated on the highest hill we had yet ascended. They consist of a lofty and massive stone structure, in which is a square tank full of cool bright water—forty-five feet deep, we were told, and it certainly looks it. On one side of the square tank the water trickles in over rough stones with a soft gurgling sound that is pleasant to sit and listen to after the climb. The aqueduct which carries this stream to its destination is borne on massive arched pillars from hill to hill until it reaches Cintra, from whence is its source. In this season the supply is scanty—terribly so when one thinks of the population it supplies. We were told, however, that a new water company is bringing a large supply through iron pipes from one of the distant tributaries of the Tagus, which will be an immense boon to the city when completed. We looked into the English churchyard with its rows of tombs and dark cypress trees, and then winding through narrow-walled lanes with villas on each side got down to the landing-stage, into the galley, and on board again at 5.30, well satisfied with our day's work. During our absence from the ship that necessary but abominable infliction of coaling had taken place, and we found a general stir and bustle, scrubbing and getting the ship clean again. Mr. Pritchett, the able illustrator of these pages had also arrived, having missed us at Cowes by just half-an-hour. He had speedily taken ship, however, in the R.M.SS. *Neva*, and we were all pleased to welcome him among us and to feel that our party was now complete. Before leaving the shore we made arrangements with Franco, our courier, for the whole party to start for Cintra on the following Monday, the 16th; so we passed a quiet Saturday on board, the crew busy holystoning decks, scrubbing boats and sail covers, bulwarks and sides—a hard day's work; but coaling always entails this, and the crew worked with a will, the children all "helping," as they chose to call it, the two boys barefooted, with trousers tucked up, and the girls with

India-rubber boots on, busy with the scrubbing-brush and squeegee.

On Sunday morning the 15th a final wash-down was given, and then when clean awnings were up, with snow-white decks, yards square, and everything taut, the ship looked as bright and clean as if coaling were unknown to her. After breakfast the children hunted up every available chair and camp stool, collected the books, and erected a temporary reading-desk. At 10.30 the pipes called the crew to muster, decked out in their best clothes and white straw hats; a fine array of manly-looking fellows they were — all veterans who had served Her Majesty, and in receipt of pensions for life. Muster over, the bell tolled for service, and when all were seated, our chaplain in his white surplice came up and our beautiful Church service was performed. There is always something especially impressive about a service at sea; the ship rests quietly on the ocean, the red cross of St. George floats over the taffrail, the sailors' deep voices join in the responses and the hymns, the canopy of heaven is above, the deep below. An excellent inaugural sermon closed our first act of public worship, and a quiet afternoon and evening brought our Sunday to a close. 15th Aug.

## CHAPTER II.

### *CINTRA—MONTSERRAT AND COLLARES.*

16th Aug. ON Monday 16th August at 2 P.M. we put Max and Willy into the cutter with our four servants and the luggage, and started them off to the shore, where they found a kind of covered and curtained waggonette waiting for them, and away they went to Cintra. At 2.30 the rest of our party followed in the galley, including Miss Power, Wetherall, and Pritchett, and got into two carriages drawn by small, but good tough horses, and started from the quay at 3 o'clock all looking forward to our inland trip. On the box of our carriage sat Franco, ready to answer questions and point out objects of interest. All was delightfully new to us: the vehicles we met—waggons on two wheels, built of solid wood and on wooden axles, which groaned and screeched under the weight of their loads, drawn by pairs of sleek oxen, and driven by stalwart peasants, a light goad in hand, fine independent-looking fellows, apparently well to do and happy; numbers of women carrying heavy loads on their heads, or driving well laden asses before them; again waggons drawn by pairs of mules—all was life and busy industry on this hilly, good, but very dusty road to Cintra. The manly, contented, and respectful peasantry are evidently a busy and hard-working lot of people, and are in themselves a complete refutation to the untravelled prejudice which dubs the people of Portugal as an indolent, lazy race.

At this time of the year the country was a good deal dried up, yellow fields of stubble, with here and there fields of stunted Indian corn—evidently a second crop; a few small vineyards, lemon, orange, and olive groves, everything coated with fine dust which the wind carries for miles in light clouds. On the roadside you pass frequent straggling villages indicating a thickly populated land, with circular hardened spaces on which Indian corn, beans, or peas were being lazily trodden out by oxen, some of which, my wife remarked, were muzzled; sometimes a mule or two, or asses ridden by boys. In some of the villages you see a church and villa. The fences are either aloe or the most rickety of stone walls. The district between Lisbon and Cintra is one of evidently very poor light soil, the rock crops out continually with patches of heather or bracken, and the road is either always up or down hill, and makes us wonder how we are to ascend 2,000 feet above the level of the sea, when we seem no sooner to have gained something than we lose it again. At last, in spite of all, we reach the last ascent and enter a long shady lane, and thickly timbered grounds, at the end of our sixteen mile drive, and driving on to the higher part of the little town of Cintra reach the "Victor" Hotel. The house is divided into three blocks, in the centre there is a sort of terrace which faces the west, and from thence a most beautiful view is obtained; just opposite is the old Palacio Real partly built by the Moors, beyond the country dotted with villages could be seen the broad blue Atlantic studded with many a lateen-sailed fishing boat, and busy steamers, the latter leaving a streak of dark smoke behind them, soiling the azure sky. It was a glorious scene, and one that our eyes perfectly gloated on after our long and dusty drive. The announcement of dinner turned our thoughts to more material joys, and after this was finished night shut out all hopes of seeing any more. Over the evening smoke Franco was consulted as to the next day's programme, and it was decided to make an early start on donkeys for the Palacio da Pena, which belongs to the King Consort, Don Fernando—father of the present king. All arrangements being completed, we went off to bed, but not to sleep, for the donkeys made the most abominable braying all night long; could we have understood them, as did our old friend in the *Arabian Nights*, we should no doubt have been amused; very likely they saw us arrive, and marked with anything but delight the handsome, not to say portly, proportions of some of the party, and were now describing them to the others, who bewailed their lot as they "bitterly thought of the morrow."

17th Aug. When the elders of the party came down at 8 A.M. they found that the children had been up long before in their eagerness for a start, and that Pritchett also had been out sketching in the cool morning air. Breakfast was soon over, and then a scramble took place for what each thought the best donkey. The ladies of the party sat completely sideways, without any pommel, on wooden saddles covered with cloth; and away went the motley and noisy cavalcade, nine in number, with shouts of laughter. The children led the way, which was at first through narrow walled lanes with overhanging creepers, branches of oleander, and other flowering shrubs, then out upon a hill side wooded with cork-oak, fir, eucalyptus-globulus, cypress, and other varieties of trees. The country was broken up in the most varied manner; sometimes we descended and crossed little dells, then rounded huge overhanging cliffs, catching occasional peeps of the fairy castle on the peak above us, its sister heights capped with ruined castles and castellated

THE CASTLE, CINTRA.

walls, in olden times the eyrie strongholds of the conquering Moor. Looking down from these wild rocky crags upon the wooded country below, and the dark blue sea beyond, one understands why the Moor struggled so long, and clung so hard, to retain this fair country, ceding it only step by step, until at last the crescent went down, as it ever must, before the cross.

View from Palacio da Pena.

Having walked round the castle battlements, and gazed long upon the sunlit scene beneath us, we passed out through the portcullis gate hewn out of the massive rock, and by paths through the woods, ornamented with cascades and fountains, pools with water lilies and gold fish, through wild cliffs, and along old Moorish walls, to another ruined castle, from which we saw the same view with the addition of the fairy castle we had just left, and which peeped out most unexpectedly in sight wherever we went. By the side of the paths they say there are planted 22,000 camellia trees—a gorgeous sight in the months of April and May; of course we were too late to see them in flower. Very reluctantly we left this scene of enchantment, and mounting the donkeys, rode back to the Victor. These donkeys are indispensable, everybody rides them, and most excellent animals they are. The roads are good, but so steep that anything like a *walk* is out of the question. In the afternoon we drove towards the village of Collares, passing Montserrat, which belongs to Mr. Cook—as he is in England—although here he is Marquis of Montserrat. It is a beautiful place, and kept in perfect order; we rambled about the woods and gardens—flowers in abundance—Cape jessamine, bougainvilla, floriopondias, bushes of heliotrope, enormous palms and tree ferns, fountains and lakes—it is impossible to describe the luxurious beauty of the scene. One could not but think of the owner escaping from London, east wind and fogs, to this haven of rest and beauty. We then descended into the fruitful valley of Collares, amidst the vineyards which produce the wine bearing its name, and through orchards of oranges, lemons,

peaches, pears and figs, fields of melons and Indian corn, until we reached the village. Here there was a sort of festa going on; a small lake had been formed by damming back a rivulet, on this there was a boat with villagers, some were playing upon the zither the soft but somewhat melancholy airs of the country, while others sang, and the oars kept time to the music; on benches under the trees there were groups of people drinking from glasses the red wine which stood on rustic tables before them, and smoking the inevitable cigarette. In front of the little shops were piles of peaches and other fruit, with patient donkeys waiting to have their panniers filled, and flags hung lazily across the street. The scene was full of colour and charm, and one felt inclined to contrast it with our own country, and say with a sigh—"Poor labourer at home, fourteen shillings a week, and the work-house before you. Happy peasant in sunny Portugal!" We looked on contentedly whilst the horses were being baited, and tasted the pleasant Collares wine, and eat fruit at what seemed to us a ridiculously low price, and then drove home to Cintra in the cool of the evening, getting peeps every now and then of our favourite castle on the Pena, and above, of ruin-crowned crags and towering rocks. We decided that on the next day we would make an earlier start so as to avoid the mid-day heat, and at 5 A.M. on the 18th, after a 18th Aug.
hasty meal of fruit—specially delicious the cool green figs—we all (with the exception of Mrs. Lambert, who was rather knocked up with the previous day's hard work,) mounted our sturdy friends the donkeys again, and started for the Convento de S^a. Cruz. We soon left the good road for a stony bridle path over a wild tract of gorse, bracken, and heather, but with flourishing young plantations. After

about four miles we got to the convent, or rather a hermitage of Franciscan friars. It has lately been purchased by Mr. Cook, who has endeavoured to restore the ruins and keep the surroundings in good order. When Franco told us we could dismount we could hardly believe we had arrived at our destination, or that this was the place in which thirteen monks used to live; we rode in between large flat rocks, which formed a rough natural inclosure, and having got off the donkeys, entered a little flat space, in the centre of which stood a cork-tree having on two sides large overhanging cliffs, under one of which there was a semicircular bench of stone; aided by the branches of the tree the overhanging roof made a pleasant shade to sit under. "But surely, Franco," we said, "this is not the convent?" "Oh, no," he answered, "I have just rung the bell," and he pointed to a door which we had not noticed, made of cork, and situated between the two large rocks, under one of which we were sitting. When the door was opened, we passed through to another open space, which was the hermit's refectory. Here again the roof was formed by the cliff; underneath were two stone oblong tables, with stone seats, between them a streamlet trickled, cool and sweet, into a square tank scooped out of the solid rock; sitting here we looked out westward, over the fertile valley of Collares to the dark blue Atlantic. The hermits had evidently retained their love for the beauties of nature when they made this refectory. The cells are some of them roofed, some cut into the hill-side—wretched places, certainly, but in this climate no doubt the monks slept for nine months in the year under the vault of heaven. We reached Cintra at ten o'clock by another road, from which we got a glimpse of the palace and

convent of Mafra, and beyond it the lines of Torres Vedras. We made an idle afternoon, lounging about the town of Cintra, the market-place and old Palacio Real, one of the rooms of which has its walls painted with magpies. There is a story of the queen of that day having ordered these magpies to be painted as a continual warning to lovers, she having been attracted to this quarter of the palace by the chattering of one of these birds, and discovered one of the royal family making love to a maid of honour. We settled to send the servants and luggage back to Lisbon next morning and join the ship ourselves again in the evening; so at half-past five on the 19th we got into two carriages for our 19th Aug.
final excursion to Mafra. This is about three leagues north of Cintra on a good, but, as usual, very hilly road. We went first to the hotel and then to the palace. It was built by John V. in the wonderfully short space of thirteen years, and was completed and inaugurated with extravagant pomp in October, 1730. It is said to be as a single building, one of the largest in the world, and barring the Escurial, probably the greatest monument of man's folly, for, situated on this lonely spot, seven leagues from Lisbon, and utterly unused, it must necessarily before many years are past, become a ruin. It is a magnificent structure of stone, containing some fine marble monuments, a library well furnished with books, a refectory, grand staircases, and two towers, each containing a peal of bells. If situated in a large town, it would have made a fine picture-gallery and museum, but where it is, there is no one to listen to the tuneful bells, to pace the corridors, to banquet in the halls, or worship in the beautiful Basilica. Sad to think it is nothing but a folly, and like all follies must end in ruin. We rested for an hour or two at

at four in the morning, we sighted Desiertas Islands, and at seven unfeathered the screw again and steamed on into Funchal, anchoring close inside of Loo Castle, in eleven fathoms of water. We almost immediately received a parcel of papers from home, and a magnificent basket of flowers from Mr. Farber, a partner in Messrs. Blandy's house. In the afternoon we went ashore in a boat engaged for the ship's use during our stay here, as the ship's boats are not suitable for beaching on the steep stony beach. The native boats are made with three keels, which keep the boat upright; when taking the beach they are turned sternwise to the shore and a rope attached to a yoke of little island oxen, who run you up high and dry in no time. The children started off for a ramble in the town, whilst we went to call on Mr. Farber and our consul Mr. Hayward, to ask advice as to what was best to be seen. In the evening as we sat on deck, we were all amused at seeing a man-of-war come into the roads by the bright moonlight, for as she got nearer, such a din of orders was heard from fore and aft, alow and aloft, as set us guessing what country she could be from. "Portuguese," said one, "nó one else could keep up such a chattering." "I don't know," volunteered another opinion, "I think she's French." Keeping our ears on the strain, presently an order was distinctly heard in English. Was ever such a row heard on board an English man-of-war? None of us would allow that this was possible, and next morning we found we were right, and that the ship was the U.S. corvette, *Saratoga;* excuses no doubt must be made for her, as she was a training ship.

24th Aug. To-day we laid in stocks of fresh vegetables and sheep; and found the latter much superior to the Lisbon mutton; there the carcases only weighed thirty-two pounds, but

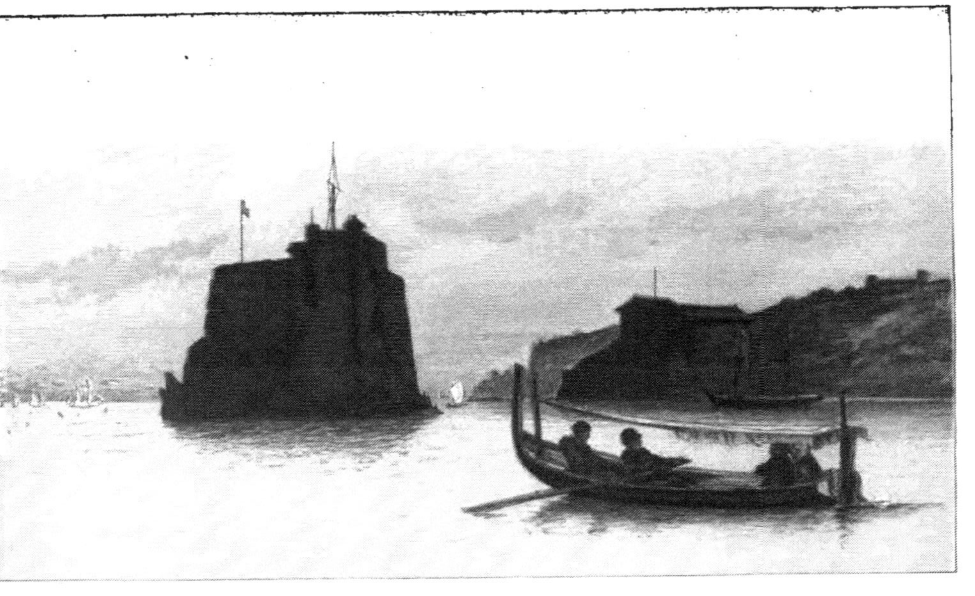

THE LOO FORT, FUNCHAL

*these* weighed from fifty to sixty pounds. We strolled into a splendid vineyard of muscat grapes, and bought a large basket at about sixpence a pound. The children came off delighted with the slide on which they had been, from the Mount to the town, and assured us the sleighs come down at the rate of fourteen miles an hour. Amongst other purchases we made was that of a grey parrot, who lived but

VIGIA, MADEIRA.

a short time. He was, however, replaced by another got at Gaboon whose history can be told in a few words. He was with us all the voyage, and came back to Cowes, where he found another home. The sailors taught him to speak, and he proved an apt scholar, for one day when Lady Brassey was on board, she went to look at him, and on the covering of his cage being removed, he eyed her with that peculiarly deep look that a parrot has, and then said in a very clear voice, "Well, my handsome!" After this it was decided that he should make his home with the lady for whom he had shown such a proper appreciation.

27th Aug. We were off in our Madeira boat at seven, and on landing found Mr. Farber waiting for us with two of the island sleighs drawn by oxen. We went to see the residence of the late Mr. Davis, with beautiful gardens overlooking the bay; the wealth and luxuriance of flower and foliage is impossible to describe—plants and flowers familiar to us only in hot-houses in England, growing here in the open air in profusion. Some of the walks lead to balustraded terraces, from which you look straight down into the sea beneath you, so clear and transparent that the rocks and stones, fathoms deep at the bottom, are plainly visible. We spent a pleasant time here, and on getting back to the ship for breakfast found the steward returned from market, and all ready for our start again. We said good-bye to our kind friend Mr. Farber, and at 11 o'clock steamed ahead. At 4 P.M. we feathered the screw, set all fore and aft sails, and proceeded with a light wind on the port quarter and a smooth sea.

## CHAPTER III.

### *FROM MADEIRA TO GABOON.*

On the 28th we sighted the Peak of Teneriffe with 28th Aug.
sufficient distinctness to enable Pritchett to sketch it, and between 3 and 4 P.M. the Grand Canary was also seen. The weather was perfect, with smooth sea and soft balmy air, and all the next day, with a nice breeze on our port quarter, 29th Aug.
we slid lazily on our way. The trade wind was very light, but on the 30th we got a fresher breeze and spun along 30th Aug.
gaily, having set port, topmast, top gallant and lower studding sails. No thudding of the screw now—nothing but the bubbling splashing sound of the ship cleaving the water. The 31st the same sort of weather, with shoals of flying fish 31st Aug.
rushing along in front of the bows. About 7 P.M. a message came from the men forward, asking leave to bury their "dead horse." This being granted, a procession was formed on the forecastle, headed by a band, composed of a piccolo and an accordion, a big drum made of the largest mess tin, smaller drums of the same style, beaten with *muffled* spoons; for want of instruments some of the other men carried camp stools, which they opened and shut, keeping beautiful time, although not producing much music; then came two men drawing after them a board on which lay the "dead horse,"

heavy rain squalls, dirty and very disagreeable up to 11.30 A.M. when it began to clear up, and the sun came out; land also was visible, or rather the trees upon it, being the coast of Liberia, which is a republic of American negroes—from all accounts a disagreeable lot. It possesses a black Episcopalian
11th Sept. bishop, ruling over a black clergy. After sunset the wind and sea went down, and we had a fine quiet night, succeeded by a pleasant morning with a nice S.S.W. breeze, so that we were able to add the mainsail to the canvas already up. It looked as if it were going to keep steady weather, so we turned the fires down, and at 2 P.M. again feathered the
12th Sept. screw; at sunset, however, the wind began to fail us, and our pace fell off from seven knots to four or five. On the 12th, which was a Sunday, we had a very beautiful day with fresh pleasant breeze, going steadily at from seven to eight knots, a bright sun, and all on board cheery and happy; we were able to have service, and found that the choir, owing to the practice, had improved very much, some of the men having capital voices. The afternoon kept calm and sunny, and all sat on deck at night in the moonlight, and we thought we
13th Sept. were in for another spell of fine weather; but about 1 A.M. next morning the rushing of feet overhead and sharp words of command clearly told those in bed below that a change had come over the moonlit scene we had so reluctantly left only three hours ago. On going upon deck, a black lowering cloud was seen coming down upon us, the phosphorescent crests of the waves rushing before it clearly showing that wind as well as rain was close at hand. Before it struck us, however, fore top-sail and top gallant halyards had been let go, and the respective sails clewed up and furled, the main and mizen throat halyards had also been let go, and as

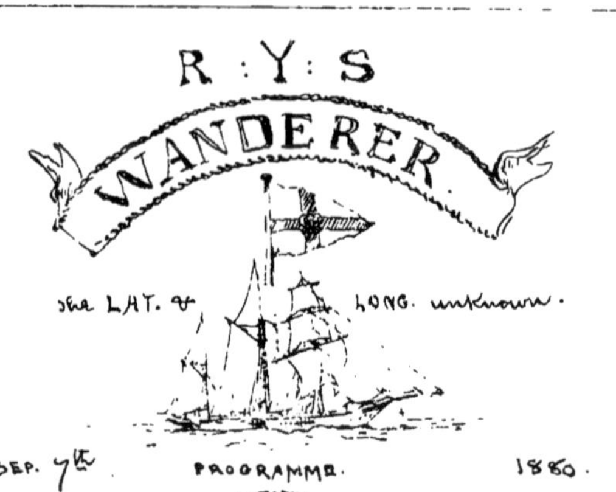


SEP. 7th PROGRAMME. 1880.

| No. | | |
|---|---|---|
| 1 | RULE BRITANNIA - | The whole company. |
| 2 | "Il balen" | Revd. H. Wetherall. |
| 3 | Hearts of Oak - | W. Gee. |
| 4 | Napoleon's farewell | R. Bowell. |
| 5 | The midshipmite. - | Young minstrels - |
| 6 | Solo on Tin Whistle | W. Hollis - |
| 7 | Frozen child. | W. Riddols. |
| 8 | Edinburgh's Town. | Miss. Power. |
| 9 | Death of Nelson. | E. Davey. |
| 10 | The Wanderer's alphabet. | R. White. |
| 11 | Our Sailors on the Sea. | J. Badge. |
| 12 | Pom-Pom. | J. Hurst. |
| 13 | Jacks every inch a Sailor - | J. Turner. |
| 14 | "GOD · SAVE · THE · QUEEN" by the whole company. including Sailors. | |

Doors open at 6. Performance commences at 6.30

VIVAT · REGINA

Under the Patronage & by kind permission of

Mr. & Mrs. C. J. LAMBERT.

the wind struck us, heeling the ship over for a while, we had nothing on her but the boom foresail and staysail. It lasted about ten minutes, and was followed by a tremendous downpour of rain for an hour or more. We had a succession of rain squalls and gusts of wind, but no satisfactory breeze all day. We were now about 700 miles from the Gaboon, and not a little anxious to get on, for the live stock was represented by one solitary sheep, the fruit and vegetables all gone with the exception of a few potatoes; but we were anxious not to resort more to steam than possible, not only for the increase of comfort got under sail, but also as it was uncertain what supply of coal we might get at the Gaboon, and we had a good bit of steaming to do from Gaboon southwards, to make a fair S.E. trade across to Saint Helena and Bahia. The 14th being 14th Sept. one of the children's birthdays, we had all the officers to dine with us, and considering that with the exception of a little fresh mutton all the dishes were composed of preserved meats, fish, and vegetables, we decided that Louis had distinguished himself, not forgetting to mention the baker, who in cakes, tarts, &c., had equally deserved well of his country.

After dinner was over we adjourned to the quarter-deck, which had been gaily decorated, and a concert was given, of which herewith a copy of the programme. All went off famously; it was a bright, calm moonlight night, very suitable for the performance, although I am afraid some grumblers among the company were wishing for a spanking breeze to enable us to make a little better progress. On the 15th we had a little more wind again, and got on well; 15th & 16th Sept. on the 16th we had a series of heavy squalls and rain,

ending in a calm, so we began again to prepare for steaming. At 2 P.M. San Antonio Point was abreast and the screw
17th Sept. set going, the weather squally and rainy. On the 17th we made land at 9.30 A.M., and anchored at Gaboon at 2 P.M., hopes of fresh milk, fruit, vegetables, meat, &c., filling our minds. Captain Gordon, Pritchett, and the two boys started off at once to see what could be got, but came back very crestfallen, a goat, two or three fowls, and a few eggs being the result of their expedition—but we hoped for better luck next day. Captain Gordon, who had been trading on this coast for twelve years, on going ashore recognised many old acquaintances. He was busy all day trying to procure fresh provisions, coal, and water, and found two friends in an old negro trader called Mackay, and a chief or king of the district, a real old heathen named Retiga, a lover of rum, and a man of many wives, through whom he is related to the kings on both sides of the river for a considerable distance into the interior. The French doctor of the settlement came off to see us as soon as we anchored, and told us the place was fairly healthy, but advised us not to be out between 10 A.M. and 3 P.M., and not at all at night. He also told us that if we landed about half past five in the morning at Libreville (the French settlement) we should find the market held under a mango
18th Sept. tree close to the landing-pier. Punctually next morning three or four of us went ashore, and sure enough there were the natives under the tree. Hurrah for the luscious fruits and dainty meats of the Gaboon! When we got up to them, "Oh, what a fall was there, my countrymen,"—a few black men, women, and children squatted together; in front of one of them three eggs (not so large as a good pigeon's

GABOON.

egg), before another half a bunch of bananas, the next had one or two yams, then some cabbages, a basket of lettuce and radishes. This was the Gaboon market! Before, however, we got over our disappointment, it was clear to us that we had better take what we could get, for the stewards of a French corvette, the *Talisman*, lying in the bay, were making short work of the supply. They were, however, very civil, offering to share with us what they had bought. I found that they had spent about 5*s.*, we managed to spend 10*s.* more—these two sums represented in full the value of Libreville market. There being nothing more to buy, we strolled off towards the habitations of the natives; as luck would have it, in our ramble we came upon a large canoe party from Cape Lopez, laden with dried and salted fish, a few alligator pears, and some chickens, and eggs. We secured as much of these as we could for the money we had about us; not a penny more would they trust us with, or even take the trouble of taking the provisions on board our ship, for which we offered to pay liberally. These fellows are well grown and good natured, but careless and lazy to a degree. As we went back we called on the Commandant who received us very kindly, offering to supply our wants in water, coal, and flour out of the French government stores. The flour and water we gladly accepted, but declined the coal, having found that a trader just arrived could supply us with French patent fuel. The poor Commandant gave us a melancholy account of the present state and future prospects of the "Colonie du Gabon." Trade, which at one time had been very prosperous and remunerative, after the abolition of the slave trade had fallen to the barter of worthless goods, old condemned Tower muskets, powder and ball, slop clothing, cotton goods, pocket-handkerchiefs, wideawake hats, copper and brass rods—the former of which are used for bracelets and anklets for the women—lastly, but above all and before all, rum, the destroyer of the aborigines all over the world. The wretched system of giving the negro traders credit in goods for which the equivalent value in produce is seldom if ever returned, has come to yield a bare pittance, in no way compensating the white man for the miserable life he has to lead, to say nothing of the loss in health brought about by residence in this hot-bed of fever and ague. One has but to look on the wan, sallow faces of the white population, traders—French officers or sailors—and notice their lassitude and utter want of energy, to determine what a fatal place this is for Europeans. The trading for ivory and ebony has not many years more of existence, the elephants are being rapidly destroyed, and the ebony tree, which is very slow in growth, is being cut down—large and small, and no young trees coming on; the only element for the future is the oil palm, which sows itself. The black man is going also very fast; contact with Europeans seems to make these people effeminate, lazy, and drunkards, in spite of the self-sacrificing efforts of the missionaries, Roman Catholics, and American Protestants, who strive to educate and civilise the youth of the country (all hope of taming the adult having been given up); yet with all this, as they grow up, a terrible percentage relapse to the vices and evil habits of their fathers.

At Gaboon there are two missions, one under the auspices of the French Government at Libreville, consisting of two establishments. In one the native lads are taught to read and write, and a trade to the more intelligent; the other is occupied by from six to eight sisters of the Sacré Cœur, who educate the girls, teaching them to work and embroider, as well as to garden. The only vegetables we could get at Gaboon came from their garden. Excellent, self-denying women are these French sisters, you find them all over the new and the old world, dedicating their lives to the good of others, teaching, and nursing, always busy, cheery, contented,

GABOON.

and loved by all. The older mission is on the site of an old slave depot or baracoon at Glasstown, about two miles higher up the river, and is conducted by American Presbyterians. Dr. Walker, who is at the head, is a highly intelligent and pleasant person; he is assisted by two ladies and another gentleman. Doctor Walker is a doctor of medicine, and his knowledge helps him a great deal to gain influence over the natives. Some years ago he left the coast with the intention of settling in America, but hearing that sickness and death were playing sad havoc amongst the brethren, and the work being much disorganised, he came back again, bringing with him Miss Cameron, a very charming young lady, for whom one could not but feel sorry, but who assured us she was very happy and liked her work, Mrs. Bushnell, the widow of a gentleman who had been previously a missionary on the coast; and Mr. Marling, a young man, but one who was seemingly very earnest in his work. We exchanged several visits with these good people; they breakfasted and lunched with us on board, and we all breakfasted at the mission with them one day, when they made the boys and girls sing hymns to us in English and Mpongue, their native tongue. They seemed intelligent children, but we were told they were in most cases of very perverse disposition, and difficult to rule. The natives who visited the ship would sell anything. Beatrice bought a necklace, which one of them was wearing, Wetherall a paddle, and Max another, leaving them only one to get to shore with.

The chief Retiga paid us many visits, and furnished a good deal of amusement. The first time he came he was dressed, no doubt, in his best clothes: from his hips downwards he wore a white satin skirt with blue stripes, and round his shoulders a mantle something like a bath towel, which he now and then removed to let in the cool air, and to give himself what Joe Gargery, in *Great Expectations*, calls "a rounder." He went grunting about the ship making his observations, and when Mrs. Lambert was pointed out to him he said, "Mammy, you no 'fraid of big water?" When she said "No," he replied that *he* was "much afraid. Ship go down, you all be drowned." It evidently puzzled him very much to understand *why* we should have come such a distance, bringing the "piccaninnies." We happened to have a mechanical bear and mouse on board; these were wound up for him, and when he saw their performances he was evidently in a great state of wonderment. "Ugh, ugh, what matter; make white man die. Who make such things? black man say bewitched." This feeling he evidently had himself, for when the bear came too near him he pulled his petticoat tightly around, and made one of his children who was with him get out of the way. The piano also was another source of wonder, but after a little persuasion he ventured to touch it. I pointed to the picture of a very pretty girl; that was standing on the table, and said that was another wife I had left at home, but the chief was not to be taken in with that, he laughed and said, "You not have two wives!" He owned himself to having fifteen, "and as many more as he can get, and twenty-four children." When he was leaving the ship he asked if he might have the "little doggie," as he called the bear, to show his people. We told him one of the boys should bring it ashore next day, and he was very particular that it should come to his house first, saying, "Me see it first, me proud." Next day Willie went with the bear, and had the pleasure of exhibiting it to the wives and children of Retiga, who presided; there was a chorus of "Aw, Aw!" but whenever it came near them they skipped about and scuttled out of the way as quickly as possible. In return for our civilities the chief got up a fetish dance at night to which some of our party, including Wetherall, Pritchett, and the Doctor, went. About 150 natives were assembled in the open space, lighted up by palm-oil torches stuck into the sand, one at each corner, around which were the bamboo huts of Retiga, his wives, and slaves. The chief sat with our party on one side, to the right *stood* the women, and the men to the left, opposite was the orchestra, consisting of three tom-toms, a rude harp, a biscuit box, and a sort of whistle; as soon as the guest arrived the overture began, a rapid hammering of all the instruments with sticks, and loud blowings of the whistle. "Dat bring de people, drive away the debbil," explained the chief. Presently they began to sing and come out by turns from either side, wriggling and wobbling about, and singing what sounded like :—

> Men. "Oke ah—Oo ah Bonnee sonnee gal."
> Women. "Ay yah! Ay Gal—Peta ba—Marjah!"

This went on for about an hour, when they struck for rum. Wetherall, who learnt the tune, was presented with a tom-tom in exchange for an old hat; long before he reads this no doubt he will have given several performances amongst his friends and relations.

Retiga did his best to urge them on to livelier efforts, telling them that King Jowa was coming, and that a certain chair was left vacant for him, but I am afraid they knew as well as we did, that King Jowa who lived two miles off was old and infirm, and hadn't the slightest intention of exposing himself to the night dews. We found next morning that we were rather in disgrace with our kind friends at the mission for having sanctioned this performance, it being one of the things they are striving to put down.

On landing one morning we had the satisfaction of seeing one of the Fan cannibals from inland, who had come down to barter ivory. He was a curious but unpleasant-looking individual, with jaws protruded like a gorilla, and teeth filed to a sharp point; we asked him, by the help of Retiga, if he would not like to eat our captain, who is decidedly comfortable in his proportions, but he answered, "No, that white man was not good, he was fetish." It was very difficult and tedious work getting anything like a stock of fresh provisions —one day we managed to buy a bullock, but he was like everything else, *in the bush;* however, Pritchett and Wetherall volunteered to go and stalk him, taking the butcher with them, and were successful in finding and shooting him, and eventually with the help of some natives we got him on board minus his head and skin, taken by Retiga as his right. I don't know what the judges of fat stock at Islington would have thought of our beef, the carcase weighed 3½ cwt. Coaling went on very slowly, as the fuel had to be brought off in canoes carrying from two to three tons each, but at last we got all we wanted and began to clear up and get the ship clean again. A day or two before leaving we had a picnic on shore, leaving only Mr. Williamson, four men, and the head steward on board. We made for the southern side of the river, and ran the boats up on a lovely beach, landing with our gear and provisions about 8.30 A.M. The men soon pitched two tents, and began to get breakfast. The steaks from the little bullock roasted on wood embers were excellent. After breakfast we had some shooting at a target, and the men went into the bush to see what they could get, but although the trees resounded with song from birds of all kinds, very little was bagged, owing to the height of the trees and the boggy nature of the ground below, covered with mangroves. Later in the day we received a visit from King William, the black lord (under the French)

of a large territory on the southern side. His Majesty, a nice-looking young negro, pupil of the Jesuit fathers at Libreville, spoke French perfectly, and behaved in a courtly manner. He was dressed in coat and trousers, hat and boots, and appeared a civilized man. One of his slaves brought with him a complete skeleton of a very old male gorilla with a splendid head, which his Majesty made me a present of, and which now stands in the hall at 29, Park Lane. The King drank ale and brandy in moderation, and went off seemingly well pleased with his visit, taking two dozen of rum with him, and assuring us that he would send some of his subjects next day with goats, sheep, fowls, and eggs, but as nothing of the sort arrived, I am afraid that the rum must have put these small details out of his head. About half past four we struck the tents and returned to the ship. One of Retiga's children had been with us on this picnic, and it was funny to see the little black creature, who was called "December," digging on the beach and being instructed in the building of castles by Beatrice. Retiga looked on at the building of the sand castles, until he was obliged to ask "*why* the piccaninnies did that," whether he thought the children were seeking for gold, or doing some very fetish business, it is impossible to say, but no doubt our doings were a constant source of wonder to
25th Sept. him. On the 25th we were back at the routine work of cleaning decks and arms and getting ready again for sea. Captain Gordon had employed McKay the old trader to go up the river and see what he could collect for us, taking about $300 worth of trade goods with him, he arrived this day in his native boat bringing thirteen goats, a goodly stock of eggs, and a very few fowls as the result of his forage up the

Gaboon. We rigged up a bamboo pen for the goats, glad to get anything in the shape of fresh provisions. Next morning the 26th, Retiga and McKay came to pay us a farewell visit; both were very grateful for the presents given them. The first got a good, old peajacket, the second an old mackintosh. There is nothing these people are fonder of than presents of clothes one has worn, as they are then sure of the quality and that they are not getting "slop" which the traders bring to barter with them. Retiga made me promise that if I came back to Gaboon I would bring him a cocked hat, so that he might eclipse his brother chiefs on state occasions. King William has a magnificent crown, a coat, epaulettes, and sword, sent, he says, to his father by Queen Victoria as a token of her appreciation of his loyalty in putting an end to the slave traffic in his dominions. For sporting purposes the Gaboon would be a good place for yachtsmen to go to. A good steam launch and a covered pinnace should be taken, both of shallow draft, to sleep in, and with reasonable care, and the aid of quinine, elephants, buffalo, leopard, gorillas, snakes of all sizes, deer of various kind, and birds of every description of plumage might be bagged, not to mention the negro wild and cannibal. Years of discretion and a family, however, do not encourage such sports. At 7 A.M. the anchor was aweigh and we steamed out of the river glad that we had seen this bit of equatorial Africa, with its artless natives, its flowers, and waving forests, but glad also to leave it, with its ignorance and its misery, behind us.

26th Sept.

26th Sept.

## CHAPTER IV.

### *SAINT HELENA AND BAHIA.*

By ten o'clock we were out of the river, shaping our course to the westward with beautiful weather. On the
27th Sept. 27th we sighted Annobon, and soon after noon crossed the bow of an American whaler, the *Sunbeam*, of New Bedford, she was busy boiling down blubber, with two hands at her main and one at her fore-top-gallant mast-heads on the look-out for further spoil, which they would not be long in coming across, as we had passed two whales earlier in the morning. By the evening we were abreast of Annobon, the southernmost of the chain of volcanic mountains which begins with Fernando Po in the Gulf of Biafra. It is a curious, timber-clad mountain, rising to a height of 8,260 feet above the level of the sea, and is said to have been a favourite place for slavers to water at in the old days. Indeed, it is not so very long ago that H.M.S. *Brisk* came down on a ship with 700 slaves, taking in water, so making a precious and most unexpected prize. For the next four days we continued having fine, bright
2nd Oct. weather, but in the afternoon of October 2nd got some roughish weather, with heavy squalls at night, and a heavy

beam sea, the ship behaving beautifully. On the 3rd, at 4.15 P.M., we sighted St. Helena, and by 10 P.M. were under the lee of the island, steaming dead slow to and fro to wait for daylight. The next morning by half-past six we were at anchor in twenty fathoms of water, close to H.M.S. *Flirt*, which was re-fitting after a cruise on the West coast of Africa. Our anchor was hardly down before a boat boarded us bringing a kind message from Captain Brackenbury offering his services. After breakfast I went ashore with the two boys to call on the governor and gather what was best to see during our short stay at St. Helena. Having given our orders for carriages and horses for the next day, we made our way back to the *Wanderer*, calling on Captain Brackenbury *en route*. During our absence on shore the French transport, *Rhin*, had come in and brought up close astern of us, having had a tedious voyage of eighty-four days from Melbourne, where she had been on special service, taking goods, &c., to represent French produce and manufactures at the Exhibition. In the afternoon we all went off for a walk in the streets of James Town, and were delighted with the contrast offered to Gaboon, our last place of call—not to mention the facilities for getting provisions. Excellent beef at 1*s*. 1*d*. per pound, very good mutton, ducks, fowls, butter, made our eyes twinkle, after having been reduced to preserved provisions, relieved only by the tough and skinny Gaboon goats' meat and wretched little chickens.

3rd Oct.

4th Oct.

We were up early, and on shore by 9.30 A.M., the party consisting of Mrs. Lambert, Captain Gordon, and myself in one carriage, Miss Power, Beatrice, Dr. Grey, and Willy in another,

5th Oct.

and Nelly, Pritchett, Wetherall, and Max on horseback. The road was up a zig-zag cut into the massive lava cliff-side. We stopped when we got to the Ladder Battery and Signal Station, and all dismounting looked out from the battlements, 700 feet above the sea, on Little James Town and the shipping, with the *Flirt* and *Wanderer* almost under our feet. We had never had such a bird's-eye view of our

JAMES TOWN, ST. HELENA.

floating home before, and probably never would again. The view over the sea to the horizon was very extensive; the signal-man said he had seen us on the evening we sighted the island, about thirty miles off the harbour. After a short rest we went on again, still climbing up, up, along the ridge that runs behind the Battery. As we ascended, we left the bare lava rock and came to prickly-pear bushes, and then to aloes in full blossom, some of them fifteen feet high, and lastly, when we reached the tower-capped height known as High Knoll, we found fir-trees, gorse in full bloom, the Australian willow one mass of pale-yellow bloom at this season, and pasture with sheep and donkeys feeding. We met islanders of both sexes and ages, and every possible shade of complexion and variety of feature—sharp or flat nosed, woolly or straight haired—driving donkeys laden with fire-wood for the consumption of James Town. Plantation House, the Governor's residence, is a massive, square building (like many that one sees in England) standing in a pretty valley, and before it an extensive plot of ground suggestive of cricket or lawn-tennis. Leaving this valley you pass a church and many a little farm, and climbing along the steep sides of overhanging cliffs, clad with timber, the fir and willow predominating, you reach the open and less precipitous parts, rich with luxuriant grass. We caught glimpses of wild, high peaks, whenever the driving mist and rain which was gradually drenching us lifted sufficiently to give an idea of the grand spectacle we were losing. Never mind!—in spite of wind and rain we were all thoroughly enjoying ourselves, and having reached Sandy Bay ridge, were driving along the high back-bone of the island, when the driver stopped and told us that we were now over Sandy Bay, on the southern side of the island, and if it were not for rain and fog, could look down on the bay, and on every peak and ridge in the island, and, out to the east, west, north, and south, over the broad South Atlantic. *Very likely!* but we had to take all this for granted, for having waited patiently for the clouds to lift, if only for a peep, until hope grew faint and fainter, and our own sodden condition became more and more a decided fact, we turned and took the road to Longwood. As we got a little lower down, a lift in the clouds gave us a glimpse of Diana Peak, 2,700 feet high; and as we approached the plateau on which Longwood stands, we found ourselves on a well-turfed stretch of land, a chance that the dripping riders had been looking out for, and away they went at the best pace they could screw their horses up to until they pulled up at the fence inclosing the House. We were received by a pleasant, civil woman in charge, widow of a French serjeant-major; it did not take us long to look through the wretched little rooms which, for the last six years of his life, had been tenanted by the great Napoleon. We moralised, as I suppose all do who visit this place, but our various ideas and thoughts shall not be inflicted on our friends. Certainly we do things better now and are more civilised than in the days that we condemned the greatest general of modern times to exile and imprisonment in the bleakest, wildest spot in this island. We then walked down to the grave where Napoleon's body lay until removed to a more fitting resting-place, "dans cette belle France que j'ai si bien aimé." In three-quarters of an hour we got down again and on board, after a day sadly marred by the weather, but with subject-matter for thought and conversation for many days.

6th Oct. This was a quiet day, in the morning we had a visit from the Governor and his secretary, and from the Commandant, Colonel Phillips, who very kindly gave me a beautiful skull of a hippopotamus (a familiar object now to my friends at home) which was shot on the Congo. Lieutenant Ross of the *Flirt* came to lunch with us, and in the evening Captain Brackenbury and Lieutenant Bennett of the same ship dined with us.

Captain Brackenbury brought his bell-band on board with him, he has trained them and leads them himself; they played old familiar English airs very prettily, and we had a most pleasant evening. Next day, our blue-jackets, who had had 7th Oct.
general leave during our stay at St. Helena, having all come aboard, our sea stock taken in, and bills paid, at 10.30 A.M. we weighed anchor and steamed under the stern of the *Flirt*, which presented an animated appearance, with her drum-and-fife band on the poop deck, the officers all on the bridge, and signals flying "A pleasant voyage" and "Farewell," dipping our ensign as we passed her, and with waving of hands and handkerchiefs we went on our way. We had not steamed more than a mile before the screw was unfeathered and all sail set, but the wind was so light that at noon we were barely eight miles from our anchorage. At sunset we saw the last of St. Helena with its rude and frowning exterior, but inland there are so many beautiful smiling spots, and its inhabitants are so kind and genial that we could take nothing but pleasant memories away with us. At noon on the 8th we had only run ninety-seven miles, giving an average speed of four miles an hour, the worst day's work since we left England; but who could grumble, with sea so blue and calm, with balmy breezes and cloudless sky; existence alone was perfect pleasure. Next day, 9th Oct.
the 9th, with the same sort of weather, we made another short day's work; at night, in the brilliant moonlight, the spars of the main and mizzen masts seemed to run up into the clear, star-illumined sky to twice their real height. On the next 10th, 11th,
two days we continued creeping along, but on the 12th it 12th Oct.
fell dead calm, and the ship tumbled about so much that I 13th, 14th,
ordered steam up and kept the engines going until the 15th, 15th Oct.

when, as there was a promise of wind, we took to the sails again, and on the 16th and 17th we got a fair breeze, bowling the ship along under square and studding-sails. At day-break a sail was sighted on our starboard bow, coming down with the wind, and every stitch of canvas up to her skysails driving her along at a bounding pace. She was a fine barque and seemed full of emigrants, probably bound for New Zealand or Australia; as she crossed our bow she hoisted the red colours of old England in answer to the cross of St. George we sent up, then came her number, which, unfortunately, was not in Lloyd's *Register*, so no doubt she was a new ship. We then gave our number, and, afterwards, by request, our longitude, to which her ensign was dipped in token of "Thank you." We were soon out of sight of each other.
19th Oct. On the 19th, the wind having fallen off to a dead calm, steam was got up again, and on the 20th, at 4 P.M., we anchored in Bahia Bay. The approach is very imposing; the expanse of bay studded with islands, towers, convents, and country-houses standing on the brow of the high land facing the sea, is quite magnificent. We anchored in the man-of-war ground to save being bothered by the Custom-House authorities, and received a visit almost immediately from Mr. Wilson, the agent of the Pacific Steam Navigation Company, who was an old friend of Captain Gordon. Having listened to all our wants he went off and very soon after sent us the latest English papers, which gave us, however, little news except the promise of a good harvest. At six o'clock next morning, I took the four children with me ashore, to inspect the market. They were highly delighted with the numbers of birds twittering, chirping and screeching, and with the monkeys mouthing and jibber-

ing, all exposed for sale. Afterwards we went to the upper town, coming down by a very fine lift, constructed by Bruuel some years ago, and back to breakfast. Later on Captain Gordon, Pritchett, and I went back to the shore to make arrangements for seeing something of the country. Mr. Duder, a friend of Gordon's, was exceedingly kind, and insisted on our all dining with him in the evening. Mr. Wilson joined us, and these gentlemen took us to the Exchange, which is a fine building, and also to see some Bahian rough diamonds; I tried to buy a few as mementoes of the place, but the merchants would only sell in large parcels. They also introduced us to the enterprising director and contractor for a line of railway from Cachoeira, Mr. Hugh Wilson, whose kindness to us during our stay at Bahia we can never forget. He was so determined to take us all to see the new line, show us the River San Francisco, and entertain us during the trip, that we could only accept such a generous invitation as gratefully as we could. I felt that it would give us a splendid opportunity of seeing something of the interior in a most comfortable manner, and in the company of those who, knowing the country thoroughly, could give us more information in a day than we could gather by a week's study of the guide-books. Having called on our consul, Mr. Morgan, and also arranged for a trip on the "Bonds" (as they call the tram-cars in Bahia) for the afternoon, we went back to the ship for lunch, after which the whole of our party went ashore. Here we were met by Mr. Duder, who put us in charge of a friend of his— Mr. Kup, who first took us by the lift to the upper town, where we got into the bond, drawn by mules, and away we rattled into the country. It was a magnificent drive, along a winding road, up hill and down dale, through woods in which, amongst the dark green foliage, the tawny red of the wild pine-apple caught the sun-light, over open plains dotted with flocks and herds, back again to plantations of sugar and manioc, sometimes out into the blazing sunshine, then through dark, cool glens, with every variety of cocoa-nut palm, banana, and bread-fruit to shade one, anon skirting a beautiful lake, until we arrived at a little fishing-village standing close to the sea, and overlooking a sparkling white beach, the setting to the deep blue sea beyond. We watched the scene before us with the catamarans coming in, the fishermen standing or squatting in them and the water washing over their feet, until we were reminded of the bond and the drive home. No one who visits Bahia should miss this beautiful excursion, even if they have but a day ashore, for surely in no other spot can the senses be so satisfied; the wealth of vegetation, the colours, the forms, the brilliant sky, are to be seen, not described. There was only one thing that marred our perfect enjoyment, and that was the brutal and senseless thrashing of the unfortunate mules by their negro driver; poor beasts, they seemed to be doing their utmost, but his hand was never still. We returned by a different, but equally beautiful route, getting out at a fine open space with handsome trees, called the Campo Grande, around which the English live. We called on Mr. Hugh Wilson and on his father, and Mr. Kup having also shown us his house, we arrived, a party of ten, at Mr. Duder's, our kind host. What should we think of such a colony dropping in to dine in England? Having been introduced to Mrs. Duder, a charming Brazilian lady, we sat down to a welcome and most hospitable dinner. In the evening several ladies and gentlemen joined our party, and at eleven we got back to the ship, tolerably tired, but all voluble in singing the praises of the country and its beauty, and of the kindness and hospitality of our hosts.

After another early breakfast next day, we started for shore taking the four servants. Our good friend Mr. Kup was ready for us, and as the railway station is about three miles from the landing-place we had to take a bond to get there. The road lay through a long narrow street with high warehouses on either side, containing tapioca, tobacco, sugar, and other products of this fertile land. As we were getting out of the bond our ears were greeted with shouts of laughter and songs, they came from a party of negresses returning from church. It was a very funny sight—the women were gaily dressed with turbans, bright coloured skirts, white chemises very prettily worked and contrasting with their polished ebon flesh, they wore heavy gold bracelets, and necklaces with gold tassels falling down their backs, and white satin shoes; on they went dancing, laughing, singing, balancing umbrellas on their heads, playing every sort of antic you can imagine, a curious mass of life, colour, and extravagant high spirits and good humour. Whether the sermon had been particularly long, or if it is the general practice of the coloured population of Bahia to come out of church in this way, I cannot say. Most likely the very fact of having had to exercise restraint for a certain amount of time on their spirits and tongues had been too much for them, and the reaction was what we saw. At the station we were most courteously received by the Messrs. Mawson, the elder of whom is the general manager, and the younger the superintendent of the railway, and taken to a saloon car with

23rd Oct.

two tables covered with beautiful flowers, and the sides decorated with evergreens. A second car contained beds for a lounge or siesta. We were told that the train would stop whenever we wished. Our road skirted the northern side of the bay, crossing some of its many inlets, and then struck inland, through plantations of sugar, tobacco, and manioc, over sandy tracts said to contain diamonds, then plunging into the forest opening every now and then on to grassy glades, with cattle, sheep, and farm-houses, and negroes huts abounding, until we arrived at a very fine sugar factory situated in the midst of plantations of sugar cane belonging to two young French engineers, very pleasant and remarkably good looking young men, who showed us all round, explained everything, and feasted us with the most delicious pine-apples, wine, and other good things. Getting into the train again we went on to the terminus at Alagonitas, where a capital luncheon awaited us. On our return we were met about half way by the general manager, Mr. Mawson, an enthusiastic geologist, who told us that on the shores of the bay, and close by the side of the line, there were some magnificent fossil beds. Accordingly, when we got to the place the train was stopped, and all got out, Mr. Mawson with his hammer, and Max with another. We were not long at work before a good crocodile's tooth and a fish's scale was got, with which trophies we scrambled into the train, and arrived at Bahia station between 7 and 8 A.M. Then we had to take to the bond again, and descend the lift from the town to the shore, so that it was half past nine before we got on board, all ready for bed after a delightful day.

23rd Oct. This, Saturday, was the day for which we had accepted Mr. Hugh Wilson's invitation to go with him for an excursion lasting until Monday, and at 1 P.M. he came alongside in his steamer, with his daughter, and Mr. Bilton the manager of the bank at Bahia, and his wife. We did not keep him long waiting, and were soon on board and steamed across the bay (the vessel very smartly decorated with flags), and entered the river Paraguasu, passing many beautiful islands, and wooded banks studded with the residences and plantations of the landowners, as well as monasteries—situated always on the tops of the hills to get as much fresh air as possible. The river is very beautiful, winding about with hills on each side, thick with beautiful trees; nearer the water's edge, huts, and groups of happy looking negroes with their canoes. At about six in the evening we reached the town of San Felix, which faces Cachoeira on the opposite bank. Here the river ceases to be navigable for vessels of more than 4 feet 6 inches draught, and here the two lines we were going to visit take up the traffic, the one in a north-westerly direction, and the other south-westerly, to strike the large river San Francisco at the point where it is no longer navigable in its downward course, owing to the numbers of falls and rapids. Mr. Wilson has an office and some bed-rooms at San Felix where some of our party put up, and the remainder at the house of the superintendent of the line, an Austrian gentleman named Kesner; here we had an excellent dinner, after which our kind host toasted us in most complimentary speeches, to which we replied as well as we could. The view from the house across the river is exquisite. Mr. Kesner had two large aviaries outside, one was full of small birds, the other contained parrots, one of which talked, laughed, and imitated dogs very amusingly. We had to make an early start at 7 A.M. on the following 24th Oct.

morning, and took our seats on three trollies drawn by a contractor's engine, which rattled along at a pace which seemed very alarming, considering that the line was not ballasted and consequently very rough, still as we had Mr. Kesner himself and Mr. Ross his assistant on the engine we felt reassured. The route travelled was again through the most beautiful scenery, and over a country that seems to call for cultivation, so teeming with luxuriant growth of every description that it looks as if there was no end to its fecundity and fertility. The line winds up the side of a glorious wooded ravine, by a beautiful waterfall, and past occasional patches of cultivation which increased in extent as we ascended, which we did for nearly twenty miles, until we reached the plateau on which the line will continue its way inland, and where the plantations are said to be much more important. At the end of the line we turned back again, running down gradients of 3°, and looking over the precipitous sides we could see the brawling stream beneath. It was very hot, and the cool shower-bath we found ready on our return can only be appreciated in imagination by those who know what a tropical climate is; refreshed, invigorated, and exceedingly hungry, we sat down to a most palatable luncheon.

I imagine that the rest of the afternoon was profitably spent by all, there was some talk of siestas, and more shower baths, but at six we all met again at Mr. Kesner's charming bungalow and had a very merry evening. Sundry toasts were proposed—success to our voyage, and so on, to which we heartily responded, wishing prosperity to the new railway and good dividends to the shareholders, which they ought to get, as the line runs through districts abounding in gold and other

metals, not to mention diamond fields, and with a soil of which you may truly say that if you tickle it with a hoe it will laugh with a harvest. On the last day of the expedition 25th Oct.
we were up at five, and, after a light breakfast, crossed the river in canoes—hollowed out trunks of trees, similar to what we had seen at Gaboon, and remarkably crank, to the north bank where it is spanned by a railway-bridge, connecting the two lines from Santa Anna to the San Francisco river. We found a special train waiting for us and travelled through beautiful country, but not of such a wild character as we had seen the day before, until we reached the terminus of Feira Sant Ana, where the weekly cattle fair was being held. We looked on at the busy and somewhat exciting scene which was going on around. Herds of almost wild cattle, kept together by stalwart countrymen mounted on wiry little horses and dressed in buff-leather from head to heel, fine, hardy-looking fellows, half Portuguese, half Indian I should imagine, but showing no trace of black blood—they reminded one of the Chilian Huasso, but they evidently have not his skill, for they carry no lasso, but are active and dexterous enough in galloping after a runaway bull and turning him, or failing this, they seize the tail, turn their horses sharply at right angles and with a jerk throw him heavily; when a bull has had two or three of these falls he seems perfectly amenable to reason, and slinks back to the herd with sore sides, a wiser creature. We also looked on for a few minutes at the slaughter of a few cattle, as there is nothing disagreeable or painful in the operation, only making us wish that the same merciful plan was adopted in England; they are struck with a short stiletto behind the horn, dividing I suppose, the spinal cord, and without a

groan or struggle the animal falls dead. We then threaded our way under a blazing sun along the sandy streets and through crowds of busy, noisy folk of every shade and colour —the yellow, narrow-shouldered and diminutive so-called *white* Brazilian, contrasting very poorly with the finely-built, muscular negroes and negresses. The shops contained a curious and remarkably nasty-looking collection of food, especially pork, raw, cooked, or cured, and in every shape. The less said about the pig in this country the better, unclean is decidedly not a strong enough word to describe his habits and tastes. Besides the pork, armadilloes, and other extraordinary and disgusting looking bits of meat were being sold, and eaten with relish by the crowd. Passing through bales of tobacco, tapioca, farina and arrow-root (the last three all prepared from the manioc root), by piles of sugar-cane, and other roots and vegetables unknown to us by name, we arrived at a little hotel and sat down to breakfast. It had been our intention to return to Bahia by the steamer which had brought us up the river, but on getting back to Cachoeira we found that the captain, instead of hauling out into the stream as he should have done, had remained in his berth, and there was the steamer stuck fast, and no chance of floating her until 7 P.M. It was especially provoking as we had invited a large party of those who had been entertaining us so hospitably in Bahia to dine on board the *Wanderer* at 6 P.M. However, it was very clear we should not have the pleasure of receiving them, and were lucky in being able to telegraph to Captain Gordon to do the
26th Oct. honours. Eventually we reached our ship at 1 A.M., and were able to make our kind friends tolerably comfortable, and induced them to stay with us until three in the afternoon.

At 4 P.M. we were off again, truly sorry to be leaving beautiful and hospitable Bahia, and the kind friends we had made; but in a voyage round the world one cannot linger just as one likes even with so many temptations to do so. Certainly the enterprise and energy of Mr. Hugh Wilson is exceptional; one cannot help feeling that the government of Brazil owes a debt of gratitude to him for the way he is opening up the country around Bahia and developing the mineral, agricultural, and other resources of the country, which must all tend to bring capital into the country and foster and promote new industries.

# CHAPTER V.

*RIO.—DOCTOR GUNNING.—THE EMPEROR AND EMPRESS OF BRAZIL.*

31st Oct. FROM October 27th to October 31st we had fine weather, anchoring at Rio at 8 A.M. on the latter day. We had been waiting outside for daylight to enter, but when the sun rose a beautiful spectacle lay before us; the rising sun lit up the outlines of the mountains, and gilded the tropical foliage of the islands which surround the harbour, it fell upon the frowning forts, and on the peaks of Corcovado, Tijuca, and Gavia, it brightened the spires of the churches and the convents, and flashed on the wild, fantastic peak which seems to pierce its way into the clouds, and is called by the natives "The finger of God." The day turned out hot and sultry, and we were glad that it was Sunday, to spend a quiet day on board, undisturbed by sight-seeing, and have our service as usual. In the afternoon, heavy black clouds began to gather on the mountain tops, and about 6 P.M. occasional gleams of sheet lightning prepared us for the coming of a storm. It grew nearer and nearer, bursting upon the harbour with such a downpour of tropical rain as few of us had ever seen, then came peal upon peal of deafening thunder, accompanied with flashes of blue forked lightning, illuminating the shipping, the houses on shore, and the forts with its blinding light—so vividly that we could not look at those flashes which seemed to strike the very ship. It lasted until half past ten, and cooled the air most wonderfully, giving us a pleasant night's rest, but the people said such a storm had not been experienced for many years past.

On Monday we received a visit from Dr. Gunning, of whom we had read and heard as guide, philosopher, and friend of all who visit Rio. He at once drew up a programme for us, of what we were to do and see during our visit, and began by taking us to the Botanical Gardens. We went in one of the bonds as the Gardens were about five miles off. The streets are narrow, but there are some fine squares, and the suburbs, with gardens, lovely flowers, and palms, are pleasant. The negroes here are much more civilised than those at Bahia, but not half so picturesque or interesting, and the negresses all seem to run to fat, and present the appearance in figure of a sack of flour, with a cannon ball at the top for a head. The Botanical Gardens are noted for the avenues of palms, averaging ninety feet in height, planted in the form of a cross. The walks were rather sloppy after yesterday's storm, and as it was still threatening and showery, we took to the bond, and back again to the *Wanderer* in good time. 1st Nov.

To-day we all went ashore, the ladies to visit the shops and public gardens, Captain Gordon and I to pay our respects to our minister and consul. Dr. Gunning came back to the ship to dine, and the after-dinner chat resulted in a decision to start next day for Palmeiras, where Dr. Gunning resides. Accordingly an early start was made, by the Pedro II. Railway, passing the city 2nd Nov. 3rd Nov.

abattoir on the outskirts of the town—the neighbouring fields, houses, and walls black with vultures waiting to do their scavengers' work. The line runs through a fine country, and at about 11 o'clock we arrived at the station and made our way to the hotel, and sat down to a sumptuous breakfast. Dr. Gunning most kindly put up Mrs. Lambert, myself, and Miss Power in his own house; the rest of the party were located at the hotel. After this late breakfast we went out for a stroll in the primeval forest, about which it is difficult to write, it being impossible to describe the wondrous beauty of the foliage, flowers, ferns, and trees,—the trunks festooned with hanging creepers, high up upon which wild pineapples were growing. Great steel-blue butterflies flitted about, looking like birds, and vainly pursued by the boys. One tree seemed to be covered with white flowers, but on examination turned out to have the upper part of its leaf of a white colour; this is the "embauba" or "sloth tree." The trunk is hollow, and a favourite resort of ants. The sloth, climbing up the tree, goes aloft and eats the leaves for his vegetables, then descending finishes off the ants for his meat. (It seems that the sloth is not such a fool as he looks.) Some of the party had the satisfaction of disturbing a rattlesnake and hearing his rattle, which is something like the clicking of a telegraphic battery, and also another snake about four feet long, of a brilliant green colour. This snake was afterwards described by one of the party as "twelve yards long and as thick as a man's leg, that he slipped through some gentlemen's legs, and then sat coiled up in a bush, sticking his head out and staring at the gentlemen." After dinner in the evening, Dr. Gunning, seeing that Mrs. Lambert and the girls were evidently somewhat tired with their sight-seeing, asked us to spend a quiet day on the morrow under his hospitable roof, which invitation all were glad to accept; and we spent a delicious lounging day sitting in the veranda 4th Nov. strolling through the woods, the boys catching butterflies and beetles, and trying to do the same to the humming-birds, but failing, I am glad to say. Sitting in the veranda these little creatures flitted about, plunging their beaks into the dark orange-coloured flowers of a creeper that grew on the side of the house. The look-out from here was on to palms, pomegranates, jessamine, roses, lilies, gladiolas, star-of-the-night, in fact, on to all the wonders of tropical vegetation, improved and cultivated by human skill and knowledge. Here also was the "traveller tree," from which, when a knife is stuck into it, water runs out; we tried for ourselves and got a glass full. It has a spike of red and yellow blossom, and the seeds have a bright blue covering exactly as if wrapped in blue silk. And so the eye travelled over many a strange and beautiful object to wooded valleys, and to hills rising behind hills in the distance. At night the fire-flies were splendid, some with a brilliant green, others with a flashing light. We ended the day with a most cheery dinner, the doctor having some old port of which he is justly proud. It was most pleasant to see him, thirty years' absence from Scotland having not in the least destroyed his affection for his old home, and with his Scottish prejudices, likes and dislikes as strong as if he had left his country but yesterday, and as cheery and light-hearted as if he were but sixteen.

We were off at 6 A.M. next day by the train to Entre 5th Nov. Rios, *en route* for Petropolis. The line runs along the banks of the beautiful river Parahiba, and through coffee and sugar plantations. At Bana de Piratic we changed trains, getting to Entre Rios about twelve. We had secured places on the coach, which was waiting—drawn by five mules, three leaders and two shaft—and taking our places started on a very fine road, which took us down the river Parahiba for a few miles, until we took the river Piabanha, along the banks of which, now on one side then on the other, we continued ascending. The scenery was wild and beautiful—giant trees, brilliant flowers, radiant birds, towering granite cliffs, and the brawling river. When we were coming to the end of the first stage, Mrs. Lambert espied four white mules awaiting us in the distance, and immediately announced that she was sure those were the bad mules mentioned by Lady Brassey. Sure enough, when a start was made, away they went anywhere but on the road, but fortunately over a fairly level bit of grass; a very few yards off, however, some gigantic boulders were to be seen, and if the coachman had not managed to pull up before we reached them, a smash would have been inevitable. All the gentlemen of the party jumped off from the top of the coach with amazing dexterity, assuring the ladies that there was no danger, and that the safest thing to do was to sit still, which they did. The gentlemen have not to this day succeeded in explaining to the ladies in a satisfactory manner *why* they jumped off, if it was safer to sit still. At one part of the road we heard a loud whistle like that of a railway, but as there could be nothing of the sort within hearing, we had to appeal to the coachman, a stolid German. He explained that it was made by an animal; but what animal? The linguist of the party was appealed to, who, having said something to the driver, after some little hesitation informed us that the coachman said the noise was made by a flying guinea-pig. It is unnecessary to say that the linguist went down very low in our opinions after this. The entrance to Petropolis is charming, with the Emperor's palace and other large houses with tropical gardens, while through the centre of most of the wide streets flow banked-in tributaries of the river we had been ascending. We pulled up at the Hotel Ingles, not sorry to get off the coach after the long drive, have our dinner and go to bed. We found that we were glad of a blanket at night, and to sit round a log fire during the next day, which turned out drizzly and 6th Nov. cold. We had meant to have made an excursion in the forest, but considering the state of the weather thought it more prudent to let the young people do this for us, which they did, mounted on horses and mules of all shapes and sizes, while we contented ourselves with strolling about the town. We had a visit from Father Vaughan, an English Roman Catholic priest travelling on church business; he acts as private secretary to Cardinal Manning, and had been through the centre of South America, roughing it in many ways, sometimes walking, and at other times carried by Indians. We tried to make up for lost time, and started off next 7th Nov. morning, some on foot and some on horseback, to a waterfall about five miles out of the town, the day bright and hot, but luckily the path lay for the greater part through forest. After fording the river, up the banks of which we ascended, we came to a double cascade, which pouring over massive granite has worn two large basins, by the side of which we might all have been seen in different positions dipping our heads to cool. After luncheon we made a fresh start on horseback to Alto del Emperador, a peak of the Organ Mountains 3,000 feet above the sea level, from which we got a lovely view of the whole of Rio Bay, the town and islands, forts and peaks. We got back in time for dinner, after which some of the gentlemen of the party went to a public ball—the resort of all Petropolis—a truly German institution, attended by the grannies and grand-papas, married couples, youths and maidens, even down to the babies. Nearly all the company were Germans or Creole Germans. We looked on at a quadrille and then a waltz, danced by everybody except the babies; it was very orderly and well conducted, but did not tempt us to stay longer, remembering that on the morrow we should be in for another long day.

8th Nov. We were up at 5.30 A.M., and after saying good-bye to our courteous and kind host, Mr. Mills, who had made us so comfortable and met all our wants with so much readiness, we got into nice light carriages drawn by four mules, and first ascended the Organ Mountains, about 400 feet above Petropolis, and then down to the plain below—3,000 feet by a zigzag road of seven miles, with constant beautiful views of the Bay and the plain beneath—where we took the train to Maua, the terminus and port; here we embarked on a very comfortable Bay steamer, arriving at Rio about 10 A.M., from whence we hurried on board our own ship to write letters for the English mail on the following day.

9th Nov. This Lord Mayor's day we spent in paying visits and in shopping; at Mr. Rickett's house we met a Senhor Nabuco, the leader of the Liberal party in the Brazilian Chamber of Deputies, a very well-informed and pleasant man.

10th Nov. As arranged the day before, I went off in the galley this morning to bring our minister, Mr. Ford, and party to lunch, after which I took them in the same boat under canvas to the steamer for Petropolis; on getting back to the *Wanderer*, Dr. Gunning announced that there was just time for us to see Tijuca. So off we went again to shore, first looking in at "Portuguese Joe's" shop, which is a general rendezvous for foreigners, and thence getting into a passing bond drove to the foot of the hill, where some of the party got into a waggonette drawn by four mules, whilst the girls and the rest of the party rode. The road was a repetition of similar beauties to those we had already seen, and though delightful to those who saw them, must become tiresome to those who can only read of them. Suffice it then to say that through wooded hills and lovely dales, amidst palms and ferns and flowers, we ascended the cliff until we arrived at White's hotel, where were assembled in the veranda the English who come up to escape from the heat of Rio. After dinner we had a magnificent drive down to Rio, watching the lights of the town twinkling below us; and taking the bond again at the foot of the hill, got back to the *Wanderer* about half-past eleven.

I went early ashore to meet Dr. Gunning by appointment, (11th Nov.) and we drove off in a tramcar to San Cristoval to pay our respects to the Emperor. We were shown into a waiting-room, and presently the Emperor came in and received us most kindly and with a perfect absence of ceremony. He promised to visit the *Wanderer* next day between twelve and one, after inspecting one of the forts. Much pleased with the interview, we returned to Rio, visited two of the ministers for whom I had letters of introduction, and looked in at the Chambers of Senators and Deputies, which were sitting, and then to lunch on board, after which we came ashore again,

and mounting horses that were waiting for us, set off for a ride up to Corcovado. We got to within a few hundred yards of the summit, when we took to our feet, and found it a pretty stiff bit of climbing, but reached the top about half an hour before sunset, and were well repaid by a most glorious view of the town—the numerous islands in Rio Bay, and the sun sinking into his ocean bed. We returned by a different road through the forest by moonlight, with fire-flies flitting round us, reaching the hotel at 9 P.M., and after a capital dinner got back again on board after a good day's work. There had been an increase to the family, the last of the Gaboon goats, who had only just escaped being turned into mutton, having presented us with a pair of kids—small goats, not gloves. The little yacht *Falcon*, which we had seen at Bahia, had also arrived. She is of 33 tons, and is what sailors would call a *wholesome* little craft, manned by four gentlemen and a cabin boy. As we had not had time
12th Nov. to visit her when at Bahia, I went to call next day, and found only one of the gentlemen, Mr. Andrew, on board; he said they meant to go round the world in her, and that they had had so far a very pleasant voyage. On returning to the *Wanderer*, found the ship cleaner and smarter, if possible, than ever; the men in their white frocks and straw hats, officers and quartermasters in their uniforms ready to receive our imperial visitor. I donned my naval reserve uniform, not to be behindhand in doing his Majesty all the honour we could. At noon the Emperor's yacht was seen coming towards us, and as soon as she was near the men were ordered aloft to man yards. The two boys stood on each side of the gangway holding the blue-and-white silk hand-ropes, and as he came alongside in his green-and-gold barge

the Brazilian ensign broke at the main. Stepping on deck, he was taken to the quarterdeck, and the ladies, who had been practising their curtsies with great vigour, were presented. The Emperor sat chatting a little, and the officers of the ship were also presented to him; he examined every department with keen interest, taking particular notice of Sir Wm. Thomson's compasses and other novelties, and going down into the engine-room. After this we all sat down to lunch; and, having put his name down in our

EMPEROR'S YACHT, RIO.

visitors' book, "D. Pedro II." promised Beatrice (who had been encouraged by Dr. Gunning to ask for it) his photograph, and asked for our own and one of the *Wanderer*. Shortly after he went off in his barge, the men manning the yards as before, and the Brazilian flag being hauled down. He passed close to us in his yacht, bowing and waving his hand; and as he went by two Brazilian men-of-war lying in the Bay, they also manned yards and cheered. It was altogether a pretty sight; and the impression the Emperor left on us was that of an able and intellectual man, quick of apprehension, and of most kind and courteous manners. He only brought two officers with him—an aide-de-camp and the chamberlain. The former gathered, I am afraid, some slightly incorrect impressions as to the owner of the yacht, as he alluded more than once to "Senhor Milord Wanderer." In the afternoon several of the ministers and their friends paid us a visit; and in the evening Miss Power and the two girls, escorted by kind Dr. Gunning, went off to the Opera, where they saw the Emperor again, with the Empress.

13th Nov. The Empress had been kind enough to express a wish to see the two girls, so this morning Miss Power took them ashore, where they were met by Dr. Gunning, who escorted them to the palace; here they were most kindly received, the Emperor and Empress giving them their photographs and autographs. We had a good many visitors during the day, and many kind friends to say good-bye to, but above all to Dr. Gunning, who had made our visit so pleasant and successful; he had never spared himself any trouble or fatigue, and seemed, during our visit, as if he only lived to make us acquainted with the beauties of the country and to make our stay a happy one. Kindly, hearty, and generous friend, he stands out with marked individuality amongst so many whom we met on our voyage. It was a perfect evening as we slowly steamed out of Rio, the rays of the setting sun casting a rosy glow over the scenery of the lovely bay, and on the peaceful sea,—a tranquil, restful evening, impressing the mind with a deep sense of gratitude and of the beauty around us, an evening to stir fond recollections of dear absent ones at home.

## CHAPTER VI.

### *MONTE VIDEO.—MR. SHENNAN AND NEGRETTI.*

OUR voyage down to Monte Video was marked by no special incident, except that we experienced some very changeable and uncertain weather, with a head wind and sea, which delayed us considerably, so that it was not until 8 P.M. of the 19th that we anchored off Monte Video. The day had been beautiful as we steamed up the Plate, with a bright sun and crisp air; locusts, moths, bees, and flies dropped on the deck, and long streamers of fine cobweb came floating on the wind until the whole of the rigging was glistening with them; some of these threads, six feet long, blew straight out, looking like blown glass. It is called "virgin's thread," and is said to come before a storm. Sure enough, next morning we found ourselves in the midst of a "Pampero," as the gales are called from blowing over the Pampas, and the flights of insects, the brilliant sunset, and the glistening threads were only signs of the coming storm. It was a complete change from the tropical weather we had been revelling in at Rio, the wind blowing hard and cold. We had an early visit from Lieutenant Dean Pitt, of H.M.S. *Garnet*, with Captain Loftus Jones's

14th to 19th Nov.

20th Nov.

compliments and offers of assistance; besides the *Garnet* there were two gun-boats, the *Forward* and *Swallow*, and several men-of-war of other nations, lying here. It was a nasty sort of day to pay visits, but, as there was no sign of improving weather, at two in the afternoon the galley was lowered, and taking Pritchett and Wetherall, with the two boys, we made for shore under canvas, through spray and breaking waves, the boat behaving admirably. As a torrent of rain was falling not much was seen of Monte Video, so having paid our respects to the English minister and had our names put down at the club, we came back to our boat, taking the *Garnet* on our way.

Sunday, 21st Nov. We had our usual service, the day bright and fresh, and could hear the band playing on board a large French man-of-war during mass. Soon after twelve Captain Jones and Commander Henderson came on board and stayed to lunch. Their boat's crew found an old shipmate or two amongst our men, so they dined on board. In the afternoon we thought it wise to take advantage of the fine weather and all went on board the *Garnet* to tea, and found one or two old acquaintances amongst the officers. It was amusing to see the men scrutinising our boat when she came alongside the *Garnet;* it being Sunday there were not many on duty, and she had over 200 on board.

22nd Nov. We did not get off next day until 2 P.M., as we had to wait for our bill of health, but at last, having taken a pilot on board (a Mr. Evans, a very efficient and pleasant man), we wound our way through the hundred miles of shoal and sandbank between Monte Video and Buenos Ayres, arriving
23rd Nov. at seven o'clock in the morning, anchoring four and a half miles off the town, which lying on low land does not present

a very impressive appearance from the sea. Soon after breakfast the usual visit of ceremony was paid to our acting minister, Mr. Egerton, and consul, Captain Mallett, and we took the opportunity of securing rooms at the Hotel de la Paix. Next day, Mr. Norton, the manager of Messrs. Lamport and 24th Nov.
Holt's business (owners of a line of steamers running from England to nearly all the Brazilian ports and those of the River Plate), came on board with Mrs. Norton to lunch, and kindly offered to take us back in their steam-tender, and save us the ducking that was inevitable if we had used our own small boats. The tide in the river was very low, always the case when the wind blows down stream for any time, and we had to leave the tender some distance from the mole, and get into small boats, finally being transferred from them into common carts without springs, which brought us through a quarter of a mile of shallow water on to a hard sandy beach from whence we walked to the Hotel de la Paix. The landing is certainly not a pleasant affair for ladies, for the getting from the *Wanderer* to the steam-tender involved a good deal of scrambling and jumping and hauling; the waves had a rise and fall of about six feet, so that those who were not quick in getting from the ladder to the boat found themselves with a fall or rise of a couple of yards, circumstances under which it is impossible to be very particular, and the sailors, with the utmost desire to be kind and helpful, handed the ladies about somewhat as if they were bales of wool. We found the Hotel a large and pretentious looking building, which is about all one can say for it. Cards from our old friend Sir Horace Rumbold and from Mr. Egerton had been left during the day, and from Mr. Woodgate, a merchant of Buenos Ayres. After dinner we went to bed, fully intending to

have a good night's rest; suffice it to say that in this we were disappointed.

26th Nov. On coming down to breakfast we found that Mr. Shennan, to whom we had brought a letter of introduction, had been kind enough to come all the way from his residence, Negretti, 125 miles off, to arrange for our paying him a visit, which was settled for the 27th, giving us a couple of days to look about Buenos Ayres and its suburbs. Mr. Shennan is a Scotch gentleman, who originally went out to Australia, but having heard of the fertility of the plains of Buenos Ayres, came to judge for himself, and being convinced of what might be done, purchased, about ten years ago, the estate of Negretti, which is six leagues square. He has done more, I should think, to improve the breed of horses, cattle, and sheep, than any one else in the whole of the Argentine province, and it is impossible to speak too warmly of his kindness and hospitality.

During the day we visited the town, which contains some very handsome public buildings, as well as private residences of the wealthy landowners and merchants; the shops also are good, but they are full of French and English goods, and destitute of any curiosities of native manufacture. The streets are most abominably paved, and a drive in a street car is a very painful style of exercise, and not free from danger, for although the private and public carriages are all made of a width, to enable them to run over the iron plating belonging to the tramcars, you have to turn off when meeting one of the cars, and then woe to your poor bones and tender places, for the streets by the side are simply heaps of stones of all shapes and sizes, with variations of ruts and holes; certainly the streets in Buenos Ayres are a disgrace to the capital of such a thriving country. In the afternoon we were driven by an Argentine gentleman, Mr. Boneo, to visit Palermo Park, lately opened; it is extensive, and laid out with ornamental ponds, caves, and rockeries, but the trees are stunted, and the place has a somewhat deserted and melancholy look about it. Our kind friend drove us back to the hotel, and after dinner we paid Mr. Woodgate a visit at his charming house in the suburbs, spending a very pleasant evening with his wife and family.

On the 26th we took tickets from the Central Station to San Isidro, a rural suburb of Buenos Ayres, to visit Mr. and Mrs. Norton; they have a charming house, cool, pleasant, and thoroughly English, a most delightful contrast to our hotel. In their garden they had a pond with a quantity of fish, which were extraordinarily tame; Willy was delighted to find that on putting his hand in they all came bobbing at it. After a very pleasant afternoon we returned to town. The railway carriages on this line are extremely dirty and very uncomfortable. 25th Nov.

We were up at half-past six, and by 8 A.M. were seated in a railway carriage with return tickets to Villa Nueva Station on the Southern Argentine line, which skirts the town near the riverside until you come to Boca, on the mouth of the river Matouzas, which is being deepened at its mouth to admit of vessels coming up to discharge their cargoes, instead of having to lie out in the boisterous river Plate, from five to ten miles from shore. Our friend, Mr. Woodgate, has this scheme much at heart, and was straining every nerve to get a Bill through Congress for the purpose, and no wonder, for at present the process of discharging a vessel is thus:—No. 1, come to anchor, if a small ship, about six, 27th Nov.

if a large mail boat, about ten, miles off, in the so-called outer roads, and discharge cargo into No. 2, a sailing lighter; as the weather permits, the lighter approaches to within a mile or mile and a half of shore, where she discharges into No. 3, a cart, which, when tolerably full, has settled down pretty well into the sand, the miserable horses standing with little more than ears and noses out of water; then ensues a fearful scene of blows and oaths until the cart is dragged out, and up to No. 4, the railway truck, with contents dripping. It is easy to imagine the time that this takes, not to mention the damage done to the cargo. Loading goes through the same process, only of course being reversed. At Rio we had seen a large Italian steamer burnt through the spontaneous combustion of her cargo, wet and damaged wool taken in at Buenos Ayres. The successful construction of a port and docks at Boca to discharge direct from ship to warehouse, and *vice versâ*, must be of enormous benefit and profit to all engaged in the trade of this flourishing country, with its thousands of square miles of grassy plains, carrying 10,000 sheep to a square league, and shearing six pounds of wool each. But to return to our journey. After leaving the town we passed through market gardens, cultivated principally by Italians, then on to vast fields of ripening corn, and lastly, out on to the prairie, a perfect ocean of waving grass, broken every now and then by a dark group of blue gums, marking the homestead of some settler. This tree, the *Eucalyptus globulus* of Australia, seems to be universally accepted as the most useful for this country of treeless plains, from its rapid growth, for shelter, and in the very near future for fencing, rafters, &c.; we passed hundreds of cattle and thousands of sheep in well-fenced paddocks;

patches of the prairie were covered with a little wild mauve-coloured flower, then white with camomiles, and again patches of brilliant scarlet verbena. At Villa Nueva we found our good host awaiting us, with a comfortable open carriage drawn by four strong horses which he was driving, this took eight comfortably, and the rest of the party followed in a dogcart drawn by a pair of horses driven tandem by his manager, Mr. St. John, and away we went, softly and pleasantly, over seven miles of grass, until we reached a good wire fence with a gate which we passed through, and were then on Mr. Shennan's estate. We drove along a road with a belt of gum-trees by the side, until we reached another gate and fence, and then through a wood, lined on each side with a fence of white iris, until we came to the open space around the house, with flowering shrubs, flower-beds, and lawns in true English style, and a most charming country house, with a large pillared veranda, built in the form of a hollow square, having an open court with a well in the centre and rooms all round. The garden was in perfect order, and every arrangement such as you expect to find in a first-class house at home, but certainly not in this part of the world, where there are no neighbours for miles. The rooms looked delightfully clean and fresh, and sitting down to tea with the open doors and windows looking out upon the garden brilliant with flowers, and breathing the air full of fragrant scents, we all felt that there certainly were other pleasures in the world besides those of steaming against a head wind and a rising sea. Mr. Shennan has acclimatised numbers of beautiful birds in his garden, he brings them from other places, and after a certain period of confinement in an aviary, lets them free; we noticed amongst others one with a crimson body, another with a primrose breast, and the "pajaros de horno" or "oven birds," which build nests of mud, in two chambers—a dwelling-room and a nursery. These birds are said never to work on a Sunday. They are reddish-brown and about the size of a thrush, and are very fond of building against the telegraph poles, which have to be protected at the tops with wire grating, or the birds' nests would interfere with the transmission of the messages.

After a cup of tea we walked out to look at the sheep-shearing, it was almost the last day's work, and had been going on for five weeks, shearing about 800 a day; the shearers come in with their families and remain until the work is done, the women working as well as the men. The stock on this Negretti estate consists of 35,000 sheep, 10,000 cattle, and 800 horses; amongst the cattle are some very fine shorthorns. There are pure bred merinos, Rambouillet, Negrettis, and English Leicesters, and a cross with these, that Mr. Shennan considers the best wool producers. We came back to the house through an avenue of sweet-scented paradise-trees, and sat in the veranda until dinner-time watching the brilliant sunset fading into tender twilight, succeeded by brilliant moonlight. The calm peace and beauty of the place, with its brilliant coloured flowers, graceful trees, trim lawns and happy birds, proving a most delicious contrast to the Hotel de la Paix.

28th Nov. Wetherall, Pritchett, and the young people were up at half-past six in the morning, and away with our kind host and Mr. St. John for a ride to a large lake covered with wild fowl, swans and geese, and with plover and snipe on its borders. They got back to breakfast enchanted with their ride over the lovely flower-decked prairie, and all they had seen, and with appetites that did justice to the Scotch scones, splendid cream, butter, and other dainties set before us. Afterwards we went for a drive over this splendid estate—the girls and boys preferred sticking to their horses—and back again to lunch. In the afternoon the boys went off to shoot with Harris, and bagged twelve and a half couple of snipe and a couple of ducks, whilst we went first to see the thoroughbred stock. There were twenty thoroughbreds driven into a corral, one of which, a very handsome black horse, Mr. Shennan had sold, and as he had to be caught a capitaz picked him out with a lasso, although the herd were tearing about the inclosure like wild creatures. The man was a picturesque object with his wide hat, poncho, gold earrings, and neat boots. We then went to have a look at the ostriches; disturbing one from her nest in the long grass, she went scuttling off down wind, flapping her wings in a fine fuss, leaving her eleven great eggs piping hot. Another one had four young ones about the size of turkeys, others with little ones that hid in the grass. It was a novel sort of farm-yard. In the afternoon we had another drive over the plains with not a stone to be seen; the mosquitoes were swarming. We saw some wild guinea-pigs which Mr. Shennan told us did a great deal of damage to the young trees, as they destroy them by eating the bark just above the earth; we went also and had a look at part of the prairie inhabited by the biscachos or prairie dogs—they are quaint-looking little animals, something like very fat rabbits, with bushy tails. We also saw alpacas, which breed and thrive here very well, guanacos, pea-fowl, and guinea-fowls.

The girls, with Max, went off again in the early morning 29th Nov.

for a farewell ride, and after lunch Mr. Shennan put us into his carriage-and-four, and let Nelly drive, to her great delight; so away we went to Villa Nueva, after a charming visit to one of the most interesting as well as beautiful properties I have ever seen. Our kind host came down to the town with us, as he had accepted our invitation to lunch on board next day, when we purposed bidding good-bye to our good friends in Buenos Ayres. Coming up in the train, I made the following notes from information supplied by Mr. Shennan, as to the capabilities of the land in the province, which are interesting, coming direct from such a practical and successful man as Mr. Shennan. Fairly good land to the west, south of Buenos Ayres, within, say a radius of thirty leagues is worth, without any improvements, about 25*s.* an acre. South and south-east of this radius, and as far as Bahia Blanca, towards which the Southern Railway is completed as far as Azul, or about half way, land is worth 5*s.* an acre. Of the first named, Mr. Shennan has six square leagues, of the second quality he has acquired fifteen square leagues in three different detached plots—Cuchico with six square leagues, La Mancha with the same number, and Oliveuse with three square leagues—which he is gradually stocking and fencing. These three estates are situated to the north of a range of low hills, a short distance from Bahia Blanca. The vast lands to the south of Bahia Blanca are in the hands of the Government, who put them up to auction with a reserve of 200*l.* per square league, or about 7½*d.* an acre. They are being rapidly taken up, if not to be occupied at once, to await the time when the railway will put them in communication with the nearest markets. The grass lands, as far as Bahia Blanca, consist of rich natural grasses, a sort of rye and clover; the seed of the latter clings to the fleeces in a most persistent manner, in fact, I think I am correct in saying that it is impossible entirely to cleanse the wool of it, and an expert will even detect its presence in the cloth after manufacture, of course detracting from the market value. South of Bahia Blanca, the rye and clover grasses begin to die out, until only tussock grass is left, which, however, is well known to our colonial sheep farmers as a very nutritious and useful grass, it covers the eastern plains right down to the Straits of Magellan.

The plains of Buenos Ayres are watered by streams taking their source from the range of hills I have mentioned to the north of Bahia Blanca; and it is a very curious feature that in their course northwards into the small lakes these streams are found to abound with fish, but when they leave these lakes and come south not a fish is to be found in them. Mr. Shennan considers that a square league of land at Negretti, and of similar quality, will carry 2,500 head of cattle or horses, and 10,000 sheep, the average yield of wool per fleece being six pounds, the best fleeces weighing as much as eighteen. The best land in Australia will only carry one-sixth of what the best land in this province can, and is on about a par with the lightest and poorest here. The chief drawbacks are the periodical droughts the country is subject to, for towards the end of a hot season the cattle and sheep get so enfeebled that when the rain sets in they die by the thousand. Last season had been a severe example, for after a remarkably long dry summer, which had been the cause of the death of large numbers, the winter or wet season commenced with a snowstorm—almost an unheard of thing—and falling upon the lean and wretched animals had absolutely killed every living creature upon many of the largest estates, and more than decimated the flocks and herds upon others. Mr. Shennan's stock was almost the only exception, he lost only about 150 head, owing to his careful and judicious management. His estate being entirely fenced and divided off into paddocks, he knows exactly what reserves of grass he has to depend upon, so that when certain paddocks on which the stock is being fed are getting nearly eaten off, they are moved on to the next reserves; and in this manner the land is never overstocked, even in the driest season, for he has always managed to have an overplus of inclosures of standing grass untouched. He also uses oil-cake for the weakly stock, which are culled and placed in a separate paddock away from the stronger ones until the critical season is over; and above all he lives on his estate, and watches over every matter himself. The estates which suffer most in other hands are those which are not fenced, consequently wherever food is to be got, there the poor beasts rush, trampling down the uninclosed districts, and eating off everything before the dry season is over, and even those that last this out can rarely stand the first cold rains. The country is wonderfully devoid of surface water, for the pools that are formed in the wet season are soon exhausted, and the beasts wander about in search of the half-stagnant lakelets, to come back again in search of distant patches of grass—a half day's journey to get a bite, and the other half to get a drink. In good and temperate seasons matters will go on well, but a bad season means destruction and ruin to the man who has not properly fenced his lands and provided for his stock. Water abounds at depths of from ten to fifteen feet, and in the carefully managed estates a well is sunk in each paddock, with a horse pump to fill the tanks, which is done morning and evening when necessary; so that in reality the difficulties and drawbacks are easily overcome in this country, which is a magnificent one for cattle and sheep breeders. With regard to the security of the settler, it appears that the governing class confine their intrigues to the capital centres, and interfere very little, if at all, with those afield, and the old days of disturbance and revolution seem to be passing away, the country settling down with a stable and reliable government. There is a steadily increasing European immigration, the best strain coming from Catalonia and the Basque provinces—a quiet hardworking people; Italians, good and frugal settlers, and Germans, Scotch, Irish, down to Russians. This medley of nationalities you see scattered about in town and country, the wild-looking but harmless Gaucho being only seen pursuing his vocation of "vaquero" in which none can rival him. While I am making these notes the train has brought us back to Buenos Ayres with its ill-paved streets, hotels, and luxuries of town life, which I am afraid none of us appreciated after our sojourn at Negretti with its flowery plains and waving groves, its shady gardens, luxurious house, and generous host.

## CHAPTER VII.

### *VALDEZ PENINSULA AND CHUPAT.*

30th Nov. BEFORE breakfast next morning we paid a visit to the wool market, now fast filling with the year's wool crop, as well as with the hides from the thousands of victims to the late snowstorm. We saw Mr. Shennan's wool amongst other, and even to the inexperienced eye its superiority was unmistakable. There was a fair proportion of good merino wool, but a great deal that was execrably bad. After breakfast we went on board the *Wanderer* in Mr. Woodgate's steam tender, our party consisting of Sir Horace Rumbold, Mr. Egerton, Mrs. Mallett and her two daughters, Mr. Shennan, Mr. and Mrs. Woodgate and their two daughters, Mr. and Mrs. Norton, Mr. Christopherson, and Mr. Bridgett, and after a very pleasant lunch bade good-bye to our kind friends, who steamed round us sounding their whistle, to which we replied with the cheerful howls of the steam siren.

1st Dec. We had a very peaceful night on our return to Monte Video, finding ourselves at anchor by dawn of day; and going ashore arranged for a special train to take our party to Florida and back on the morrow. As we came back to the *Wanderer* we called on Captain Brickdale of H.M.S. *Forward*.

2nd Dec. We started next day at nine o'clock and arrived at

Florida at one, got out and rambled about the town, and returned to the *Wanderer* in time for dinner, at which we were joined by Captain Brickdale. If we had had more time I should have been glad to have visited a Swiss colony of some 500 families, settled about seventy miles from Florida, which is said to be a very thriving community. For the defence of their families and properties they have formed a body of 500 mounted riflemen, who practise at the butts every Sunday evening. Some months before our visit to Florida a band of marauders who live by theft, and under the pretence of politics are the terror of the country, suddenly made a raid upon a herd of cattle belonging to this colony, and were driving them off, when they were overtaken by a party of the Swiss riflemen, who, on coming within shot, dismounted and fired a volley into the robbers, killing several, whilst the rest made off as hard as they could go. The Swiss quickly drove their herds back again, having given these fellows a lesson they are not likely to forget. These bands of ruffians, and the lack of justice from the government are, it is said, causing the rapid depopulation of Uruguay, which was at one time a thriving little state, for the inhabitants are flocking over to Buenos Ayres, where the soil is better, and where peace and order are to be found in a much larger degree. It is difficult to understand why this lawless little state, containing barely 400,000 inhabitants, is allowed to exist, or why we gave it up; had we continued to retain our hold on it no doubt at the present time there might have been a thriving colony of a million or two. I suppose the jealousy of the European powers is the explanation, but if so we will hope that Brazil will shortly solve the difficulty by annexing it.

3rd Dec. To-day we paid some visits in the town—to Mr. Evans, whose garden is quaintly decorated with old figure-heads of shipwrecked vessels, and to Mrs. Theobald, the wife of one of the resident merchants, to whom we had brought letters of introduction; and were glad enough to find coaling over when we came on board, and the ship reported as ready for leaving.

4th Dec. At one o'clock our kind Montevidean friends came off to a farewell lunch—our minister, Mr. Monson, two Misses Monros, very charming young ladies, Mr. Yarrow, Captain Brickdale, and Mr. and Mrs. Theobald. The latter brought Mrs. Lambert a most beautiful bouquet in the shape of a lyre, which measured three feet two inches in height: the body and stand were composed of yellow and white roses, gardenias, jessamine, and stocks, the strings of fuchsia flowers; she also had another nosegay sent her which measured five feet in circumference. Our guests left us about five o'clock, and steam being up we were soon off on our voyage again.

6th, 7th, 8th Dec. During the next three days we got some fairly fine weather, and on the 8th we were off the S.E. point of Valdez peninsula at 1 A.M., and at daylight entered New Gulf and anchored in Pyramid Bay. It was lovely weather, but decidedly hot. Coming down we were surrounded with shoals of porpoises, the beautiful clearness of the sea enabled us to see them far below racing along, then rising to breathe and down again, sometimes turning underneath the bottom of the ship. Barnes, the carpenter, harpooned one, which was five feet ten inches long, and the men breakfasted off it. Max, who tasted a bit, said it was like a fishy beefsteak. After breakfast we all went ashore, the gentlemen of our party taking their guns and luncheon in case of getting too far away from the ship; the ladies had the crew of the galley to take care of them, whilst they collected shells, of which there were great quantities, not only on the beach, but in the low hills which rise from the shore, and which are formed of a conglomerate of shells and clay. They were rather alarmed by some mysterious bellowings that they heard, and which we afterwards found out were made by sea lions, of which there were numbers in the caves along the shore. Mr. Tyacke, with a boat's crew armed with rifles, killed several, but could not succeed in bagging a single one. The party on land did not get much better luck, for the day was too hot and sultry to encourage us to wander very far away; we only shot three of the curious Patagonian hares, but saw a large herd of guanacos grazing in the distance, partridges, duck, and traces of ostriches in the shape of egg-shells and feathers, horned plover, and long-tailed screeching parrots without number. We walked inland for some three or four miles to get an idea of the soil and its capabilities, and were convinced that it is excellently suited for cattle and sheep rearing. There is plenty of tussock grass, and about the centre of the peninsula there are two lakes covered with wild fowl of every description; the water is brackish and salt, but I have no doubt that water is procurable at a very shallow depth. I was especially curious to visit this peninsula, from having often heard my father say that, when travelling in these parts some sixty years ago, he had made the acquaintance of an old Frenchman who had procured a concession from the government of the peninsula, and, having erected a fence across the neck some four miles wide, which connects it with the mainland, stocked it with cattle, and passed his life here in the company of a few gauchos, chasing guanacos and ostriches. He used to make periodical visits to Buenos Ayres to sell the produce of his estate—dried fish and beef, skins and feathers, in a wretched little schooner he owned. My father described him as a happy old ruffian; but one day he sailed from Buenos Ayres in his leaky craft with a year's stores, and was never heard of again; no doubt his rotten old tub went from under him in a pampero. Nothing but the fence across the peninsula now remains to tell of its former occupant; the cattle have doubtless fallen a prey to the Indians who occasionally visit the place in search of game, or may possibly have been driven off by the gauchos when they saw no signs of their master's return. It is capable of being made a very convenient and productive sheep or cattle run, with its natural fence on three sides of salt water—beautiful gulf and excellent harbour. Up to the present time, however, I believe the government of Buenos Ayres have refused to sell it. In the evening two of the boats with the boys went off to the shore with the seine, and were very successful, securing quantities of good fish, large and small—enough for a two days' supply for all on board. Everybody enjoyed the visit here immensely, with its perfect freedom; the men too were as jolly as possible, without the chance of spending any money.

9th Dec. At six next morning we steamed across the Gulf to Port Madryn on the opposite side, and fired a gun, on the remote possibility of its being heard by some member of the Welsh colony of Chupat, forty miles away to the south; but having concluded that this was unlikely, it was decided that a party of volunteers should start to walk on to the colony, pay it a visit, and see if horses could be sent back for the remainder of our party. Accordingly, at four in the afternoon, the little expedition started for their forty-mile walk, consisting of Pritchett in command, Doctor Grey, Flowers, the boatswain's mate, Blight, quarter-master, Hooper, A.B., and my valet, Harris, who, being a Welshman, might be useful as an

THE DREARY TRAMP TO CHUPAT.

interpreter. They were all armed in case of meeting Indians, and carried water and cold tea, bread and potted meat.

We thought it advisable that the start should be made in the afternoon, so as to avoid the midday sun, for it was really no light task—a forty-mile march, heavily weighted, over an unknown, rough country, without, as far as we knew, a drop of water on the way, requiring a good deal of stamina and *go;* and we knew that it would be impossible to hear from them before the day after to-morrow. Having seen them start, we passed the afternoon in rambling about the shore and examining some curious cabins cut out of what can hardly be called rock, it is so easily worked, being more like a sort of soap-stone. These must have been made most likely by seal-hunters—one of them had a door, window, fire-place, and cupboard, and a bed-place cut into the stone—and there were the names of many sailors and their ships.

10th Dec. In the morning we went ashore with a seining party, but only caught a few small fish—the wind blowing right in shore, and the sea getting up; so we went on board again, the bow of the pinnace being stove in as we came alongside (fortunately the galley was not injured). We found the glass falling, and the sea and wind rising fast. This, on a dead lee shore, with a drift of thirty-five miles, and a bottom that we were not at all certain of, made us determine to make all secure, and steam out for the night; so between 6 and 7 P.M., after waiting for the chance of a change, we stood across the Gulf to the weather shore, with a stiff easterly gale right in our teeth, which sent the spray flying clean over the poop deck—a thing we had not experienced before. It blew hard all night, and in all directions, until daylight, when the wind began to fall, and we returned to our anchorage.

11th Dec. About an hour before we got in we noticed a large fire far away on the road to Chupat, and we all began to wonder what was the news from our walking party. Presently another fire broke out somewhat nearer, then another and another, until one was made close to the beach, and we could see three men, but none of them belonging to our party, coming over the sand-hills leading three horses. The galley was now lowered, and taking the children we went to meet the strangers. On landing, we found them to be two Welshmen, Mr. Hughes and Mr. Jones, from Chupat, and Mr. Quick from Buenos Ayres, who had come down to visit the colony. Mr. Hughes told us that our party had arrived at Chupat at half-past two the day before, having walked the distance of, not forty miles, as we had thought it to be, but forty-five miles, in twenty-two hours—a good march for men heavily accoutred and out of training. He said they were all well, but very tired and footsore. He gave me a letter from Pritchett, who said that they had suffered a good deal from thirst. The colonists were anxious that we should pay them a visit, and had sent two vehicles to bring us over, which we had not caught sight of from the ship as Mr. Hughes had left them behind the sand-hills. All then went back to the *Wanderer* to have lunch and hold a consultation, and as we understood we could drive over, it was decided that Mrs. Lambert, Wetherall, the two girls and the boys and I should go back with Mr. Hughes and Mr. Jones, taking as little luggage as possible, but plenty of maps and food. The yacht, taking Mr. Quirk, was meanwhile to go round to the mouth of the Chupat river, where we could join her.

This being all settled, we made our start for shore again, and scrambling over the sand-hills came to where the traps were waiting. They were a kind of spring-cart, with two wheels and shifting seats, capable of carrying five besides the driver, drawn by one horse, and a dog-cart drawn by two horses driven tandem. The four children got into the first-named, and Mrs. Lambert, Wetherall, and I perched ourselves upon the dogcart. We had brought two beakers of water for the horses, which were disposed of before starting; and away we went for our forty-five mile drive—at least away went the shandrydan with one horse, but our more ambitious team had very clearly made up their minds that they were not going off without a great deal more persuasion. The leader turned round and surveyed us with inquisitive admiration. The wheeler, disgusted no doubt at not being able to see us, bucked, kicked, and backed, and the more Mr. Hughes tried to explain the case with the aid of his whip, the more did this animal refuse to understand it. The whip, I must say, was a short one, and unable to reach the leader. We had seen in our experiences between Entre Rios and Petropolis something of the difficulties of making a start, so were not very much disheartened, particularly when good Mr. Hughes was doing his utmost. A fresh arrangement of the tenants of the dog-cart was now suggested—namely, that I should come and sit next Mrs. Lambert, and take Mr. Hughes's place, and that he should *lead* the front horse whilst I drove the shaft one. Well, it wasn't exactly one's idea of driving tandem, but as we had nothing better to suggest, down got Mr. Hughes and seized the inquisitive leader, whilst I got on to the box seat and prepared to try conclusions with the wheeler. In this fashion we made our start, not at a rapid pace, for we could hardly expect Mr. Hughes to do much running, but it gave an air of great safety to the proceedings. After a little time, as matters seemed to be going on pleasantly, Mr. Hughes thought he might relax his attentions, and try a seat in the cart again—but do you think that leader was going to stand any nonsense of that sort? Not a bit of it! As soon as he felt that his friend and companion was gone he began to look for him; the loss of Mr. Hughes's company evidently occasioned him great distress of mind. He went away to look for him on the prairie; could not find him, and tried on the other side. No, still out of sight! Then he must look *behind* the cart, and when he found him sitting comfortably on the back seat, that horse was so delighted, that if there had been more room, I have no doubt he would have got in and sat by his side. It was clear that at this rate we should take considerably longer in going to Chupat than our advance guard had taken on foot, when suddenly an idea struck me—let Mr. Hughes ride the horse. The amiable beast made no objections to my proposal. Mr. Hughes got on his back, and away we went, everybody happy, until we caught up the shandrydan which had got some miles ahead of us, and another arrangement was now suggested, for the horse in the shafts had shown repeated objections to Wetherall sitting behind; in fact Mr. Hughes agreed with him that the cart was not balanced in the fashion that Chupat horses like; so we took Max up with us, and gave Wetherall the shandrydan in exchange. Mr. Hughes, too, thought that it would be possible and advisable for him to try a little driving again, for although he had had a nice ride, so far, yet the absence of saddle, and the presence of harness and buckles had begun to tell upon the outward man; accordingly up he got, and on we went in a fashion; the leader didn't like it at all, and shuffled along in a protesting style, taking very good care never to get into his collar, but not playing any remarkable pranks. The driver of the shandrydan kept on calling his horse by name to reassure him. The animal rejoiced in the singular name of "Coaken." As pronounced it sounded uncommonly like what the children said it was, namely, Cock Hen, but his master's voice, and the sound of this endearing title certainly had its effect upon Coaken.

The road was a mere track over the seemingly endless plains, the scenery most uninteresting, nothing but tussocks and little stunted bushes on all sides of us. At last

ON THE TRAMP

some miles ahead smoke was seen, and Mr. Hughes said that this was a signal made by the gaucho in charge of a relay. We answered by simply setting fire to the nearest bushes, and after a mile or two to some more, and then again and again; this being done to cheer up the gaucho with the hopes of speedily seeing us, and preventing his getting sick of waiting. About an hour after we saw the fire, the gaucho himself appeared on the plains, and by way of a diversion offered Willy a ride behind him, which was eagerly accepted, and away went the gaucho and Willy, envied, no doubt, by every one of the party, at a gallop.

About sunset we got to the fire and halting-place, and all got out to stretch our stiffened limbs and have something to eat. We had our luncheon basket with sandwiches and other things, but all preferred some excellent home-made bread and good butter, which Mr. Jones, the driver of the shandrydan, produced. We stopped until dusk, forming, we flattered ourselves, a highly picturesque group; unfortunately there was no one to look at it, and Pritchett, who could have reproduced it, was no doubt enjoying the fat of the land at Chupat. Our gaucho turned out to be a man of experiences, having served under Rosas in his youth, and we chatted away round the fire, with Coaken and the other horses grazing on the tussocks until it was necessary to make another start. Water had been left by the carts as they came for us, so we were able to give some to the horses. And now for some fresh arrangements. The gaucho who ought to have brought two horses had only brought one, so it was clear we could not do very much in the way of a change for two carts, with one horse; however, it was arranged thus—the shandrydan evidently was the cart that was entitled to the new horse, so he was put in the shafts, whilst the wheeler was promoted to be leader. Then for our own cart Mr. Hughes decided to turn out our wheeler altogether, which he did, and put Coaken in the shafts. This again was perfectly fair, for Coaken really had had a very easy and light nat

T.

plai

and

sc

sa

of

nec

th

house of the chief man in the village, and head in fact of the colony—Mr. Lewis Jones. We naturally expected that his house would be close at hand; fancy our feelings on being told that we must get into the cart again and drive a further three miles over the plain! At last, however, our journey was finished, and we found Mr. Lewis Jones most kindly

INTERIOR OF HOUSE, CHUPAT.

sitting up and expecting us. Mr. Jones had lately had his house carried away by an inundation of the river, and as his family were in England had contented himself for the present with a simple cottage made of mud bricks with four or five rooms. He had prepared two rooms for us, one for the girls, e other for Mrs. Lambert and myself, and at half past

5 A.M. we found ourselves getting into bed very tired and exceedingly glad to have finished our drive. The unfortunate horses had done over ninety miles in twenty-nine hours, but I have no doubt were as pleased at being turned into the bush as we were to turn into the comfortable beds prepared for us.

## CHAPTER VIII.

### *SANDY POINT.—ELIZABETH ISLAND AND FITZROY CHANNEL.*

On getting up next morning we found that the *Wanderer* 12th Dec. was anchored off the bar of the Chupat river, but that we should not be able to cross the bar before high water at four in the afternoon. This seemed to suit exactly, as it would give a few hours to look about, and see what progress this little Welsh colony had made since the first pioneers landed about sixteen years ago. The river mouth being seven miles from the village, we had in the first place to arrange for our party being conveyed, some in the carts and some on horseback, and also to engage the pilot boats to take us over the bar. It was decided that we should all meet at 2 P.M. in the village, so as to run no risks of missing the tide. These matters being settled, we sat down to a most hospitable and excellent breakfast, and must have astonished Mr. Jones with our appetites, and the way we cleared off the beefsteaks, eggs, and delicious bread and butter. Hardly had we finished our breakfast before Mr. Jones announced that the wind had chopped round to the eastward and was blowing fresh, and that under these circumstances it would be impossible to get over the bar.

This was not very cheering news, for I was anxious to push on into the straits as fast as possible, having determined to sacrifice our visit to the Falklands in order to get on. Accordingly we determined that we would carry out the original programme if possible, and at 1 P.M. started for the village, picking up the walking party who were waiting for us; and on we went, in spite of the assurances of the Joneses, Davises, and Hugheses that although the pilot boat had been sent down to the mouth of the river, there was not the least chance of our getting over the bar, and that it was not an uncommon occurrence to have to wait ten days or more. On reaching the mouth of the river we found the pilot boat, but the occupants told us the same story, and advised me to mount an adjoining sand-hill and see for myself. This I did, and found that they were only too correct; the sea was breaking badly on the bar from end to end; the wrecks of two vessels, one on each side of the entrance, seeming to confirm all that had been said. Two miles off, to the north-east, lay the *Wanderer* at anchor, with steam up, rolling lustily, but, alas! all communication with her was impossible, for there was not a code of signals in the colony.

It being clear that we could not leave by this tide, I thought it was possible that as the wind was generally very light off shore in the mornings we might get away with the morning tide, and this proposal was accepted as feasible; so back to the village we went again, to get some dinner, and lie down afterwards in our clothes, ready for a start. At midnight, half asleep, shivering, and, I am afraid, somewhat out of temper, we began to assemble by the light of a brilliant moon to discover fresh difficulties awaiting us. Mr. Jones had kindly provided the dog-cart to take us, but as this would not hold five I determined to walk, and as we had a good four hours to do the nine miles, this would not delay us much. The horse was speedily harnessed, and we *all* got in, as Mr. Jones, who must have had some misgivings, I think, wished to *lead* the horse at starting; and now then all are ready. Come up! Co-o-om up! Get on! Tug, whip, pat, coax. Back went the milk-white steed,

BEACON FIRE, CHUPAT.

round went the milky-white creature; nothing could have been more determined than this beast was that he would not be of the party at this unearthly hour. At last, after twenty minutes spent in useless efforts, poor Mr. Jones started off to borrow another horse, leaving us to take the harness off the obstinate one. What our feelings were, as the time quickly slipped away whilst I detached the white demon, can only be imagined. I never felt, I think, such personal feelings of ill-will to any horse as I did to this one; and as he shambled off I treated him to a kick which doubtless hurt my foot a good deal more than it did him. After about a quarter of an hour Mr. Jones appeared with another animal, which no sooner caught sight of the collar that I was holding than he first began by trying to make a clean bolt of it, but being disappointed in this, spun round and round, with occasional well intended, but fortunately ill directed, kicks at either Mr. Jones or myself. In this exciting struggle we were now joined by Mr. Jones's woman-servant, and at length succeeded in getting a noose round the creature's nose, and whilst Mr. Jones and I held on to this with all our might, the servant managed to get the collar over his head, and finally the rest of the harness on, and into the shafts, and Mr. Jones, mounting the box, we did make a start—at a very slow pace, and with occasional halts, or spasmodic rushes to one side or to the other. When we reached the village an effort was made to exchange this horse for another, but during the negotiation I walked on, thinking that if they failed I had better send on the bulk of our party, who no doubt would be waiting for us, and anyhow prevent their missing this tide. Before I reached the boat, however, the cart caught me up, and at my wish drove on, so that if in time they might get into the boat and off to the ship: fortunately, however, there was no need for this, for I got up in time, and jumping into the boat, we pushed off at once, waving our adieux to Mr. Lewis Jones, who had treated us with so much kindness and hospitality, and to whom we had given more trouble than any one, I should think, had done since the establishment of the colony. In fact all our party had received every attention and kindness at the different houses they had stayed in; but owing to our anxiety to get on with our journey, we had no opportunity of in any way returning the hospitalities and civilities we had received, and as we steamed off, feared they might think us but an ungrateful lot.

This very interesting colony was started about eighteen years ago, and I believe the original idea was that the religious sects and the native language of Wales should be strictly adhered to, and that settling down in some portion of the New World, they would live a simple, primitive life, free from contact with strangers and foreigners. Mr. Lewis Jones, Mr. Madryn, and another gentleman formed the committee, and having heard of this valley of Chupat as likely to suit their requirements, chartered a little sailing craft, and starting from Buenos Ayres, proceeded to New Gulf, a place visited only occasionally by sealers, and landing on the south-west side of the gulf in a little bay, called it Port Madryn, after one of their leaders. From thence they found their way overland, until they reached the valley and river of Chupat, where they found good soil for wheat-growing, a tolerable harbour with a practicable bar, plenty of scrub for firewood, and good tussock grass for the future flocks and herds, besides the promise of plenty of sport—guanacos, ostriches, Patagonian hares, and wild fowl of every description being plentiful. The commission accordingly returned to Buenos Ayres and obtained a charter from the Argentine government, who claim the sovereignty over this district; and returning to England, came out again in 1865, bringing about 150 immigrants, chiefly colliers, from South Wales with, in some instances, their wives and families; Mr. Michael D. Jones, the Principal of the Theological College at Bala, advancing £4,000 towards the installation of the colonists, who had a hard time of it in their weary tramp from Port Madryn to Chupat, glad, as a Mrs. Davis told us, even to suck pebbles coming over the waterless plains. The first settlers knowing little of agriculture, found themselves in 1866, when visited by H.M.S. *Triton*, almost destitute of the staff of life, and most thankful for a supply of flour and biscuits. In 1871, H.M.S. *Cracker* looked in upon them, and found they had had no communication with the outer world for thirteen months. It was not, I believe, until about 1874 that they began to avail themselves of the power of irrigating the lower lying lands; from which date their condition began to improve. The system pursued is still a very imperfect one, and Mr. Jones is unable to induce the little colony to work together and construct a canal by which an excellent scheme might easily be worked, and for which the sweat of the brow is alone required; but public spirit seems sadly wanting. With, however, the present primitive plan, the colonists have managed to get themselves pretty well clear of debt, the wheat exports bringing in a good round sum. Excellent cheese and butter are made, and they have capital beef and mutton, fowls and eggs. There are six chapels of different denominations, but the hope of keeping out a foreign tongue has had to be given up, as Spanish must be spoken to the gauchos, and in communicating with such authorities as the captain of the port and of the Custom House. Next year they are expecting a fresh importation of 500 souls, with, I hope, some young unmarried women amongst them—for when we were there, in a population of 800 there were only three marriageable maidens. To the north, south, and west of Chupat the lands are capable of carrying large herds of cattle and of sheep; for the latter, a systematic plan of burning the low thorn scrub would be necessary, as it is a fàtal enemy to wool. In fact, with increased energy there is no reason why Chupat should not become a thriving, wealthy colony, far beyond the dreams or hopes of the present colonists, with the exception of such men as Mr. Lewis Jones and a few others, who see far beyond the majority.

At five o'clock in the morning we were off, and Mrs. Lambert 13th Dec.
and the children having gone to bed, the gentlemen of the party followed their example. The next day the glass fell 14th Dec.
rapidly, and we found the weather much colder, and on the 15th at 9.15 P.M. we sighted Cape Virgin, the wind 15th Dec.
increasing in fury, the glass still falling. On the following morning the gale was so strong that with an ebb tide, we 16th Dec.
were unable to reach Cape Dungeness, and put back under Condor Cliff, on the north shore of Cape Virgin, and came to anchor, riding comfortably for the rest of the day and night. Next morning the glass showed no signs of rising, and the wind blew in tremendous gusts. Captain Gordon 17th Dec.
described the weather as a furious gale with hurricane gusts. We could do nothing but stop where we were. On the 18th 18th Dec.
the glass began to rise very slowly, giving us hopes of getting away; but the wind still blew too furiously for us to move all day, but began to moderate at night. At two next 19th Dec.
morning we were delighted to hear the cable being hove in, and at half-past twelve were again steaming at full speed past Cape Virgin, 'and were soon entering the straits with

SANDY POINT

during the day, but found only two Chilian gentlemen and one lady as first class passengers; we also paid our respects to the governor, who entertained us most hospitably. The riding party went off to visit a wild gully where boring for coal was going on, and also gold washing. Dr. Fenton dined with us in the evening, and from what he said it seems as if the colony at Sandy Point had fallen into decay. The old coal-mine is no longer worked. The railway line leading to it is impassable; burning trees in many places have fallen across it, and here and there it has been broken up by floods. The working for gold is carried on in a very desultory manner; sometimes a wandering colonist will take a few days provisions and squat up the ravine until he has washed enough to secure him a long drink on his return, but even this style of gold digging has become very rare.

At ten next morning we steamed to Elizabeth Island, about 22nd Dec.
fifteen miles from Sandy Point, and anchored in Royal Road. Mr. Reynard has stocked Elizabeth Island with sheep, and keeps a shepherd there with some horses which he very kindly placed at our disposal, and which the young people found a great resource. The island is well covered with tussock grass affording excellent pasture, and the sheep and cattle we saw were in capital condition. There is no doubt that the whole of the eastern side of Patagonia affords good grazing. We landed soon after our arrival and set off to "kill something." The grass and scrub contained quantities of wild geese, but they rose at long distances, and it was only when we surprised them on the crest of a ridge that we got any shooting. There was a brisk wind blowing which gave us a quick sail going between the shore and the ship; every now and then we crossed long masses of kelp, drawn by the wind

and tide at right angles to our course, and the shocks they caused were sometimes sharp and sudden enough to make us doubt for a moment if we had not struck upon a rock.

23rd Dec. The day following, it was decided to get as much sport out of the island as possible. We were told that in the centre of the island there was a lagoon swarming with wild swans and geese, so taking the galley's crew to carry the collapsible boat I set off with a duck gun to see what was to be got. Unfortunately it was the breeding season, and the surface of the water was covered with ducklings, goslings, and cygnets, it would have been wanton cruelty to have destroyed them or their parents, but the blue jackets picked up some goslings, thinking to rear them on board. Meanwhile the two girls had ridden across the island, coming to a place that looked as if covered with snow, but as they drew nearer clouds of white terns rose from the ground, and it was difficult to avoid treading on their eggs, they dismounted and picked up a good number, which were found to be very good to eat, in taste and appearance resembling the plover's egg. A third party, composed of Wetherall, Pritchett, Willy, and a couple of men with crowbar and pickaxe, went to explore some mounds and old heaps that are spoken of in the publications of the staff of the *Challenger*, and succeeded in digging out seals and guanaco bones, and some stone fragments with marks upon them that made it just possible for them to have been handled by men of the rudest age. The captain took the steam launch to the island of Santa Marta to explore, and the ladies roamed about the beach. We had meant to have had a picnic, and a lamb was roasted for us by the shepherd's wife, but the weather turned wet and considerably interfered with the roast, as the fire had been lighted

CORMORANT ROOKERY, ST. MARTA.

outside the shepherd's hut, and when the lamb was brought in on a long stake, one side was raw, and the other cinders, however, we managed to hack some bits off it, and were not sorry to get back to our own comforts on board ship.

The morning promised well, so some of our party started 24th Dec. to visit the islands of Santa Marta and Santa Magdalena in the steam-launch, but the two girls and Max preferred riding again. Running under the northern cliffs of Elizabeth Island, eagles, geese, and ducks gave us plenty of shooting; and on landing we found ourselves in the midst of a cormorant town. The nests are raised to about a foot above the ground, made of clay and guano worked into a hard paste, and systematically arranged in streets; the top of the nest is hollowed out like a shallow basin and lined with down and feathers. Pritchett has given a capital representation of this colony of cormorants. The birds swarmed on and about the nests by thousands, letting one come to within a few feet, and just drifting slowly off—when you seemed as if you must tread on them—with a croaking, crackling sort of noise. Looking over the edge of a cliff we saw a huge sea-lion napping on a rock in the water about thirty yards below us, we waited while Pritchett made a sketch of him, and then gave him a volley of rifle-balls; his skin was so tough, and stretched over such a mass of blubber, that he had life enough left in him to struggle off his rocky bed and wallow about in the water, and when we got round in the boat, what with the kelp, big boulders, and a long reef of rocks, we were not able to secure him. We found that Santa Magdalena was as wonderful for its show of penguins as Santa Marta had been for cormorants. As we approached the land we could see lines down the hollows in the cliffs which were at first mistaken for water-courses, but turned out to be penguin runs; and as we neared the shore the rows of these birds that had been staring at us from the top of the cliffs would run down into the sea and out again. They are the most ridiculous birds, and look at you first out of one eye and then out of the other—for their eyes are so far back on the side of their heads that they cannot look straight in front of them. In their erect position, with their little fin-like wings flapping with excitement, they looked like diminutive soldiers, and as we landed, ran back evidently to give the alarm, grunting like pigs. As we ascended the cliffs we came across a stronger reserve which in their turn fell back upon larger bodies, and accumulating in this way until they formed regular regiments waddled off inland, affording a most comical sight. We followed them up over the brow of the hill, to find, when we had got to the top, that nearly the whole army had disappeared, and wondered at first what could have become of them, until we heard all round us their funny, piggish grunts, and looking about in the grass found that the ground was riddled with burrows, in each of which were a pair of penguins. We looked into some of the holes, finding one or more young ones, most lustily defended by their parents. No doubt every one had a family, but it was not a tempting place to explore, for the stench was poisonous. On coming down to the beach we found that our blue-jackets had made a fire and cooked some capital Irish stew. Our party was joined by a Spaniard who had wandered all the way down from Buenos Ayres by land, and had settled down as shepherd in charge of some 500 very fine sheep on Magdalena, the grass being better than at Elizabeth Island. He told us he had a wife and child, but that having gone to Sandy Point to get necessaries about five months back, had not as yet been able to get back to them. On another occasion he had been nine months alone, but seemed quite happy, and said he wanted nothing but tobacco, so we gave him all we could muster between us, and half a bottle of brandy, with which he walked off to his wretched hut rejoicing. Coming home we touched at Sylvester Point, the north-eastern extremity of Elizabeth Island, where there is a tern rookery, and picked up in a few minutes more than 2,000 eggs, which were taken on board the *Wanderer*, and proved a great treat to all hands.

Christmas Day opened brilliantly with a lovely morning, 25th Dec.
but before twelve it began to rain and then to hail, with a little thunder and lightning. We had service at eleven and sang the Christmas hymns, and after this was over went forward to see the lower deck before the men sat down to dinner. They had decorations of a shrub with small red berries, and white flowers like camomile, and had the Elizabeth Island shepherd, his wife, and child as their guests. He said he had not had such a Christmas Day for many a long year, which no doubt was the fact. Poor creatures! their existence must be a miserable one; but I suppose its charm lies in its independence and idleness. The little girl enjoyed herself, too, immensely; it must have been a glimpse of quite a new life to her, having only father and mother for her sole companions. In the evening the officers dined with us, and we did not forget to drink "Absent friends." Though it was midsummer the day was not a pleasant one, with repeated heavy showers of rain and hail; and the next day was worse, 26th Dec.
with tremendous showers of hailstones as big as peas, making it very cold, and driving us off the deck.

At 6 A.M. we were off for Sandy Point again, and on 27th Dec.

arrival found that H.M.S. *Kingfisher* had taken up the position we hoped to have occupied, and was coaling, which would cause us further delay. Captain Thornton, and his first lieutenant dined with us. The day was miserably cold and wet, and we had to carry out a disagreeable duty in discharging two of our men whom we had been obliged to send on shore to be locked up before going to Elizabeth Island. They had been ashore and got drunk on the horrible alcohol of the country, and came on board to fight like demons, being separated with difficulty and secured with handcuffs to the rigging, until the captain of the port came to take them on shore. One of them was a bad character, and the other, when he had had too much, was apt to be exceedingly violent; so that both had to go.

28th Dec. The *Kingfisher* sailed for the Pacific, so we were able to take her place by the coal hulks. The weather wretchedly wet again. Dr. and Mrs. Fenton and their children came to lunch. Dr. Fenton made Mrs. Lambert a present of a very beautiful guanaco rug, the centre made from the small white fur over the heart, and the border of the usual yellow-brown colour. In the afternoon the governor mounted us, and took us for a ride into the forest, and to see some coal borings, which struck me as indicating lack of mining knowledge, and as having been worked in a somewhat haphazard style.

29th Dec. At 1 P.M. our good friends, the governor and his wife, Mr. Reynard, and Dr. and Mrs. Fenton came off to lunch and to say good-bye; all had shown us the same kindness and attention that we had received at every port we called at. At sunset the ship was washed down, and we had taken in our stock of fresh meat, hoisted up the boats and made everything ready for a start next morning. When I had been at Sandy Point some four years before our present visit, it was a penal settlement, and appeared in a fairly flourishing condition. Some persons may have thought that the convicts were made to work too hard, and were kept in with too tight a hand, but the governor was an energetic man, anxious for the prosperity of the colony, and doing all with the best intentions. The coal mine had in its early days a good number of men employed in it, and great hopes were entertained of its future. Gold had also been discovered, sheep and cattle runs started, saw-mills erected, and the seal fishery flourished. Capital and trains were no doubt sadly wanting to develop these industries, yet to the eye of a passer-by there was an air of life and progress about the place. Soon after this time, however, the little garrison, and the convicts, mutinied, and maddened with drink, became like wild beasts, killing all they met, and desolating the place. The governor escaped on horseback, others saved themselves by flying into the forest. The governor after a few days managed to communicate with a Chilian man-of-war that was surveying to the eastwards, and on the further arrival of first an American and then an English man-of-war, the rioters fled to the interior, taking some women and children with them, and leaving ruin and destruction behind them. Their fate was a wretched one, for wending their way to the north, starvation stared them in the face. The brutal leaders of the rising began to quarrel, fight, and kill each other; the miserable mothers flung away their children to die in the bush, some fell dying of hunger, others, it is said were killed by Indians; a few stragglers reached the nearest Argentine settlement, and received aid and protection, notwithstanding their well-known crimes, and in spite of the protestations of the Chilian Government. For a time the colony was in a state of collapse, but now sheep are increasing in number and improving in quality. It is a pity that the Chilian Government do not give more encouragement to sheep farmers, by granting long leases at a nominal rental, in the same way that they give mineral grants—for though they tacitly permit the occupation of the land, they will give no security to the holder. The seal fishery continues to yield good profits, and the boats engaged in the trade were all away when we were at Sandy Point. Notwithstanding its rainy climate, I think if the settlement was carefully nursed it might become a fairly prosperous colony.

Dec. We steamed off from Sandy Point at four o'clock in the morning, taking Mr. Reynard with us to investigate the loss of a Brazilian brigantine, said to be ashore in Fitzroy Channel. The morning was, fortunately, fine, enabling us to see and enjoy the scenery. In the distance were towering mountains covered with snow, nearer ones with only patches of snow, and then dark cliffs closing in the channel in the foreground. As we got under the lee of Mount Tarn, a canoe with five Indians came paddling off, one of the men had a shirt on, *the other four had not!* They had a fire burning on some sand in the bottom of their canoe, which was a very primitive vessel made of strips of roblé bark, and requiring a good deal of baling. Having given them some biscuit, they produced an empty bottle, but did not induce us to fill it with rum. The woman had a square of otter skins which she would have liked to sell, but it was not a tempting looking article, and we let her keep it, and steamed on. On the passage between Cape San Isidro and Cape Froward a grand view opened out down Magdalena Sound; snow-clouds and mists went streaming over Dawson Island, and beyond in the far distance Mount Sarmiento appeared—a mass of jagged ridges and broad spaces gleaming with light, suggestive of glaciers and snowfields, making a beautiful contrast to the black cliffs rising out of the calm waters through which we were passing. When we got into English Reach it was difficult to realise one was in an arm of the sea, the water smooth as oil, broken only by the splashing of the seals and wild fowl, and the plunging of our screw. At 7 P.M. we anchored in Borja Bay, under the great masses of granite called Thornton Peaks. The hills down to the water's edge thickly clad with timber, a waterfall rushing down a ravine and pouring into the bay close to us, and all around the sound of dripping waters. After dinner we landed, but found the walking very bad, the ground seemed composed of a thick bed of rotten vegetation, and sopping moss. There were a good many boards stuck upon the trees with the names of ships, and as it seemed the fashion, we left our card in a similar manner nailed to a tree.

We left Borja Bay at 4 A.M., and as we entered Jerome 31st Dec.
Channel had a hard struggle with the tide, the sea rushing in whirlpools, and breaking as if over shoals, in a very alarming manner; but after passing Corona Island, we were able to steam fairly ahead again. In Jerome Channel the scenery is magnificent, vegetation of the richest character descending into the water; little coves, creeks, and islands all about you, until the entrance into Otway Water, where the two mountain lines close inwards and give a passage between immense heights, sometimes of bare granite, sometimes of rock covered with a coating of variously tinted growths, streaked

and rushing with waterfalls, and capped and mottled here and there with snow. Every now and then the wind came in violent gusts, and then all around would gleam and glisten under a belt of sunshine. When well into Otway Water the mountains changed their aspect and sank into green sloping downs. Not very far from Fitzroy Channel we sighted a sail ahead beating to windward, which we concluded must be the Brazilian brigantine Mr. Reynard was in search of, and on our hoisting our ensign and asking if she needed any help, she answered "short of food," upon which we stopped, and lowering one of the boats sent her to bring off the captain. The ship proved to be the *Santos*, laden with coal from the mine at Skyring Water, not being able to get a tug to take her, she was sailing down, and missing stays, drifted under a strong current on to a shoal point, where she stuck until the next tide, when with the help of a steam anchor she had hauled off again. This channel, with its strong current running from five to six knots, is not a fit place for a sailing ship to work through, and should only be attempted with the aid of a strong tug. We were glad to find the captain did not look much the worse for the "short food," and having sent him back to his ship with a quarter of beef, a sheep, and some preserved meat, steamed into Fitzroy Channel, which has some awkward twists in it—sometimes you see no outlet until you come close to the entrance, and at one point we had to drive through a quantity of kelp. The wild swans here were a very pretty sight, they rose and flew away in front of us, with their black heads and necks and white bodies. At 5 P.M. we anchored off the coal mine settlement in Skyring Water, and after dinner went ashore to have a look at it, being very kindly received by the manager, a gentleman from Buenos Ayres. We went into the pit on the incline of the seam from its cross on the surface to a depth of fourteen yards, which is as far as it goes; from hence a gallery extends from which the coal has been taken. The seam, as might be expected so near the surface, is very broken, and the best of it is only a poor lignite; on the surface there are some good wooden houses, where the manager and staff reside. Although a good deal of money must have been expended on the undertaking, it is situated in such an out-of-the-way place, and the quality of the coal is so indifferent, that I am afraid its speedy collapse may be foretold.

## CHAPTER IX.

*OTWAY WATER.—PUERTO BUENO.—CHASM REACH.—ENGLISH NARROWS.—THE GULF OF PEÑAS.—COQUIMBO.*

1881. 1st Jan. EARLY on New Year's Day we left the coal mine anchorage and steamed back through Fitzroy Channel. The entrance from this side is not visible until close to it, and you seem as if steaming on to a bare cliff. It was bright and fine, but the wind blowing from the glaciers at the west end of Skyring Water had a sharp chill in it. We put Mr. Reynard down at the outlet into Otway Water, as he had horses waiting to take him about forty miles overland across a grassy plain which he has stocked with sheep and cattle. The eastern side of Otway Water is not yet surveyed, but high hills can be seen here and there. About ten o'clock we saw the poor *Santos* again, apparently not a bit further on her way than when we last saw her. We steamed on, and round Cape St. Jerome into Crooked Reach, passing Snow Sound with a beautiful glacier, then by Snowy Channel on the Fuegian coast with two monster glaciers, and a fourth in Glacier Bay. Rounding Thornton Peaks there was a bright gleam of sunshine, and from a spot where the rays of the sun fell upon the deck arose a curious steaming appearance, followed by the same appearance in a part which was in shadow; whilst wondering what this could arise from, fire was reported in the wine and spirit room, under the dining-cabin fire-place. It might have been a serious matter had the spirit boxes been burnt instead of those which contained claret; as it was, with the aid of the extincteurs, Mr. Tyacke the carpenter, and two blue-jackets, it was soon put out.

The weather in the afternoon got very thick and lowering, and a westerly gale blowing in from the Pacific, driving hail and rain in our teeth and shrouding the landmarks, caused us to miss Puerto Angosto where we had meant to anchor for the night, so that we had to lie on and off in Long 2nd Jan. Reach until morning, when, as soon as it was light, we drove into the teeth of the gale, and passing Cape Upright made for Cape Tamar, the rollers from the Pacific meeting us, showering us with spray, and making the *Wanderer* plunge and quiver from stem to stern.

About 9 A.M. a violent gust of wind swept the whole scene clear and gave us a fine view of the twin peaks of Cape Pillar, with the great waves of the Pacific dashing masses of foam against the rocks which form Cape Phillip on the opposite coast. Mount Joy showed out plainly too, giving us our course for the Fairway Islands; once round these low barriers of rock we could gaze at ease on the marvels of wooded rock, islands without number, snow-clad cliffs, and sparkling waterfalls which surrounded us. From the Fairways we rapidly threaded our way up Smyth's Channel, through Mayne Channel, opening out from which we had hoped to have seen Mount Burney standing out in all his glory, but although the weather was fine his highest point was hidden by a cap of cloud. Winding our way through the rocky passage between Sumner and Long

Islands we entered and anchored in Isthmus Bay at 3.30 P.M. Two or three of our party landed, taking rifles in case the natives (who had lately had some brushes with the sealers) should prove disagreeable, but none appeared, although we found the carcase of one of their huts near the log track which has been laid for taking canoes across the narrow neck of land between Oracion Bay and Union Sound. This log-way is referred to in the sailing directions as "a native production," but the timber has evidently been cut by axes with a good steel edge, and not by the stone implements used by the Fuegians.

3rd Jan. To-day we passed through the finest scenery we had seen on the whole of our travels, the magnificence and grandeur of which is not to be described. We left our anchorage at half-past six in the morning; behind us were the hills of Rennel Island and Zach Peninsula, and rounding Stanley Promontory into Victory Pass, we passed close to Brinkley Island, so up Collingwood Sound and Sarmiento Channel, with Piazzi Islands on one side of us and Carrington Island on the other, to Esperanza Island, and passing up two-thirds of its length turned down a gloomy channel with dark, rocky sides and hollows, making a striking contrast to the harbour of Puerto Bueno, with its low, green islands and smiling aspect, in which we anchored, to remain a day with the object of watering the ship from a cascade of fresh water at the head of the bay. Throughout the day the Cordillera de Sarmiento afforded us an ever-changing scene of splendour and beauty; at almost every turn a fresh aspect was presented of the most fantastic shapes imaginable—spires, cones, towers, and pinnacles thrusting themselves into the clouds, massed into mighty gleaming glaciers, giving the most extraordinary effects as the sun fell upon their green, translucent rifts and chasms.

We passed a very pleasant day in the peaceful little harbour, the children paddling about in the dingey catching plenty of fish, and collecting mussels, of which there were quantities. We landed and went up to the lake from whence the stream issues at which we were taking in our water supply. Neither Indians nor water-fowl were to be seen, although traces of the former were found in the frame-work of some huts. The lake is long and narrow, winding at the foot of a mountain; the vegetation rank and beautiful, with numbers of lovely little flowers, ferns, and mosses—some of these were white and feathery, and were almost like seaweed. About four in the afternoon, from some high ground near the lake, we saw a large iron-clad coming slowly into the bay; she must have seen our mast-heads behind the island which forms the inner anchorage, for the French tricolor was flying with a rear-admiral's flag. As our people evidently did not see her from our deck, I sent off the galley to tell them to hoist our ensign; and shortly after went to pay the stranger a visit. She proved to be the *Triomfante* going out as flag-ship in the Pacific. Her captain was most kind and polite—he had heard of us at Bahia and Sandy Point—and sent us some fresh beef, to which we replied with a basket of fish just caught. 4th Jan.

We left Puerto Bueno at 4 A.M., on a very squally morning, the wind coming in sudden and unexpected bursts. Just before entering the Guia harbour, two or three squalls struck the ship with such force, that, though the topmasts were struck and everything snug, she heeled over as if under full sail. The wind was even more boisterous where the 5th Jan.

[illegible] Channel opens out into the P[illegible] [illegible] Bay, where we saw [illegible] French [illegible] [illegible] the whole scene was [illegible] and [illegible] charged with rain. When we got under the [illegible] N[illegible] mountains the weather [illegible] [illegible] as we [illegible] for Cheeri [illegible] [illegible] had been especially raised as to the [illegible] passage, and a general [illegible] manifested itself [illegible] the ship. A [illegible] was at the wheel, the [illegible] was [illegible] the head of steam, a quarterm[illegible] [illegible] to stand by the engine-room [illegible]. The [illegible] who were off watch clustered in groups about the [illegible] the cooks put their heads out of the galley windows [illegible] servants appeared in the doorways of the deckh[illegible] knew that we were within a mile or two of the [illegible] to all appearance, when the vessel's head was [illegible] [illegible] of the [illegible] of rocks and islands [illegible] Summer Island, she was running straight at a [illegible] precipitous rocks. However, the wind was whistling [illegible] with great violence, which showed that there [illegible] [illegible] of [illegible] somewhere ahead for it to come thr[illegible] In a few minutes a black patch showed itself am[illegible] wreaths of murky mist, suggesting the mouth of a [illegible] From this point a dark body of cloud bore down on us [illegible] the whole [illegible], [illegible] us to grope our way with [illegible] and after a few moments of further anticipa[illegible] [illegible] like a horse [illegible] a jump [illegible] it was deemed safer to mak[illegible] [illegible] by way of [illegible] Sound. As we turned, th[illegible] [illegible] broke, and the sun shone out, making [illegible] as if it

3rd Jan.

Trinidad Channel opens out into the Pacific, and passing Tom Bay, where we saw our French friend lying snugly at anchor, the whole scene was blurred and dirty with a heavy gale charged with rain. When we got under the lee of the Cathedral Mountains the weather improved, to come on again thick and nasty as we stood for Chasm Reach. Our expectations had been especially raised as to the wonders of this passage, and a general curiosity manifested itself throughout the ship. A second man was at the wheel, the chief engineer was questioned as to the head of steam, a quartermaster was told off to stand by the engine-room door. The blue-jackets who were off watch clustered in groups about the forecastle, the cooks put their heads out of the galley windows, and the servants appeared in the doorways of the deck-house. We knew that we were within a mile or two of the opening, but to all appearance, when the vessel's head was inclined to the left of the tongue of reefs and islands protruding from Saumarez Island, she was running straight at a mass of precipitous rocks. However, the wind was whistling in our faces with great violence, which showed that there must be a sort of funnel somewhere ahead for it to come through, and in a few minutes a black patch showed itself smoking with wreaths of murky mist, suggesting the mouth of a tunnel. From this point a dark body of cloud bore down on us, filling the whole gorge, obliging us to grope our way with slowed engines, and after a few moments of further anticipation we saw the *Wanderer's* head swerve round like a horse refusing a jump, and found that it was deemed safer to make Port Grappler by way of Icy Sound. As we turned, the storm seemed to break, and the sun shone out, making a most lovely and brilliant rainbow which seemed as if it sprung

STRAITS OF MAGELLAN, ICY CHANNEL

from the ship's side—a perfect arch with both feet in the water. As we looked up Icy Sound, flashes of light seemed to be thrown from the surface of the water, which were explained as we drew nearer by a procession of icebergs, which at first seemed of a pure white, but as we got closer to them they were of a clear blue colour. Just as we neared the head of the procession, an opening in a wall of rocks disclosed the passage to Port Grappler, and turning sharply into this we missed a clearer view of the bergs. The harbour was a very snug one, with the usual beautiful surroundings of timber-clad hills, little green islands, and snowy peaks in the background. These little islands are so trim and neat that they might be artificial; all have a rim of stones just round the water's edge, and then the green mound rises thick into mosses and trees. The boys soon had the dingey down when we came to anchor, as Nelly had caught twelve good fish over the bulwarks.

We had made up our minds that we would not be done 6th Jan.
out of a visit to Chasm Reach altogether, there were also some very fine glaciers in Eyre Sound which we had got a peep of yesterday, and which we wished to get a nearer view of, so it was decided to give a day to exploring the neighbourhood. Unfortunately the day was again misty and showery, but a start was made at 8 A.M., rounding the northern end of Saumarez Island. At the entrance to Escape Reach we had to *slow* while a dense down-pour of rain passed over us, after which we did manage to get on and into Chasm Reach, Mrs. Lambert and the children attired in mackintosh coats and sou'-westers to face the torrents of rain. The dark faces of the cliffs rise almost perpendicularly from the water, and in some parts actually hang over the channel—on all sides they

stream, rush, and hiss with countless cascades and streams of every size and volume, from the roaring torrent to the silvery rivulet, all pouring into the silent waters beneath. Occasional openings in the crags give one a distant view of some monster snow-peak in the background, but the impression produced by this channel is one of gloom, and it was almost with feelings of awe that we threaded our way through its sombre reaches. Passing Cape Patch the weather somewhat improved, giving a hope that we might be able to carry out our programme, and have a look at the Eyre Sound glaciers and Seal Rookery; however, it only enabled us to pass our friends the icebergs of yesterday, and then down came wind and rain, shutting out everything from our view, so there was nothing to be done but to return to our anchorage in Port Grappler. In the afternoon the boys and girls went to the head of the harbour to explore, and for two or three miles up a river, half canopied over with trees. They landed in a little valley, and found a wigwam made of arched sticks, roofed with bark, the floor covered with boughs, but no signs of the owners. They brought back large bunches of fuchsia, scarlet lapigeria, a small white flower like a wood anemone, but with a delicious scent, and varieties of holly, one with a long scarlet and yellow flower, another with a flower like a pomegranate. At 6 P.M., to our surprise the French flag-ship came in and anchored, and her captain came to return our visit made at Puerto Bueno. He intended sailing early on the following morning, and going through English Narrows and so out to sea. At sundown the band of the *Triomfante* came on deck and played "God save the Queen," and we all went up and showed ourselves to let them know we appreciated the compliment. Before night all boats were hoisted in and everything made ready for a start. The night was again very squally, and heavy rain fell; and I think all had begun to get rather tired of the ceaseless rain and wind of the straits, and convinced of the wisdom of Admiral Fitzroy's advice, viz., "Put on your waterproof clothes when you enter the straits by Cape Virgin, and do not take them off again until you make your exit at the Gulf of Peñas."

At 4 A.M. the French flagship was away and we followed about an hour afterwards, in windy, drenching weather. Steaming down Indian Reach we noticed that the rocks (of which there are several just awash) were covered with sea-lions. As we passed Eden Harbour we were surprised to see our French friend lying at anchor again. From this point the weather got so thick and nasty that we had to grope along dead slow, seeing only about 100 yards ahead close to the water's edge, as we approached English Narrows. Off Port Simpson the clouds lifted and we eagerly availed ourselves of the opportunity to enter the Narrows, with its whirling currents that made our bow fly round from port to starboard and from starboard to port, before the helm could meet her; but with two good men at the wheel and full speed on, in ten minutes we had accomplished the intricate and difficult navigation of English Narrows. It was a very bad day to enjoy scenery, the rain pouring down and the temperature bitterly cold, and very soon after passing the Narrows it came on as thick as a hedge, but having now plenty of sea-room we pushed on, determined, if possible, to get out to sea before night. Passing down Messiers Channel, by Island Harbour, we were surprised to find the *Kingfisher* at anchor; she signalled to us "You are welcome here," but we resisted the invitation, and had hardly time to

7th Jan.

signal back our thanks before a violent hailstorm, accompanied with thunder and lightning, divided us from each other's view. At 9 P.M. we were fairly out into the Gulf of Peñas, and at ten o'clock were pitching in a very heavy sea from N.W. and W., which knocked the ship about tremendously. It was a most disagreeable night, the ladies and children suffering sadly. The ship rolled, plunged, and quivered throughout the night—smashes of crockery and glass resounded, accompanied with the falling of every article not properly secured. The globe of gold-fish that had hung from the ceiling in the boys' cabin since leaving Lisbon came down with a crash. The flower-jar with maidenhair in the swinging table followed suit, drawers opened and displayed their contents, chairs ran wildly about; sometimes we found ourselves involuntarily rising on our feet, to be followed by an apparent endeavour to stand on one's head, then we clutched at the sides of beds or cots, as everything seemed to slide away from us, to be thrown back, as cans, chairs, shoes, brushes, &c., &c., came clattering after you.
8th Jan. Thus we jumped and danced about until next morning at 8 A.M. when, having Cape Tresmontes on our beam, we were able to change our course and set fore and aft sails, and the good ship steadied down to her usual good behaviour, and at noon, steering our course along the west coast of America, the ladies and children began to show symptoms of
9th Jan. returning life and jollity. On the next day we got some fine, bright weather, and all hands were busy drying the soaking things. In the afternoon, our course being changed to north, with fore and afters set, we made good work, bringing our speed up to between ten and eleven
10th Jan. knots an hour without pressing the engines. On the 10th

we saw land on the starboard beam, but towards night it got very thick. Through the rain at 8.50 P.M. we sighted Mocha Island, it was such uncomfortable weather that we decided not to go into Lebu, although I much wished to see an old friend who was in the habit of visiting a property he has there about this time of the year. The harbour is open to the north-west, and, as the wind would have prevented our holding any communication with the shore, we decided to go straight on to Valparaiso. On the 11th we made good progress, and on the following day, at 3 A.M., on a glorious morning sighted land, and from this hour kept Aconcagua with his snowy cap in full view, until as we went on, the whole of the magnificent Andine range, which makes the view from the Alameda in Santiago so beautiful, was in sight. At 12.45 P.M. we made fast to a buoy in Valparaiso Roads, finding no men of war, except the Chilian transport, *Angamos*, in harbour. Directly after lunch some old friends came off to greet us and tell us the news, the most important item of which was that a decisive battle before Lima was expected to take place from hour to hour. We found that our Lebu friend was in Santiago, so that we had missed nothing by not going in there. Having just gone ashore to call for our letters, we were off again at 6 P.M. for Coquimbo, which was to be the terminus of the first stage of our voyage. About 11 P.M. a thick fog came on suddenly just as I was getting into bed, and hearing the steam-whistle blow, and feeling the speed of the ship reduced, I went on deck to see what was the cause; turning afterwards to go below, just as the ship gave a sudden lurch, my feet slipped from under me, and I came down heavily. It was some time before I could get my breath, and I felt that I must be badly hurt as I was in a

11th.

12th.

great deal of pain, but thinking it might arise from bruises only, I went back to bed; but getting no rest, and finding I was quite unable to rise in the morning, I sent for our good doctor, who very soon found out that besides bruises I had broken three ribs on the right side. This was not very

COQUIMBO ROCKS.

pleasant news to receive just as we were within a few hours of our destination; knowing also that I had much both to do and to see. However, there was nothing for it but care and patience; Dr. Gray soon strapped me up, telling me that it would be a fortnight or so before the bones would begin to join, and four or five weeks before I should be able to ride about freely. We had had a succession of fogs in the early 13th Jan. morning which caused repeated stoppages. At twelve noon we had Lengua de Vaca abeam; soon after the Tongoy smelting works, and then saw our own woods of dear old Compañia. We anchored in Coquimbo Roads between H.M.S.S. *Turquoise* and *Osprey* and the storeship *Liffey*. My son-in-law and Weir were soon on board; and as the doctor told me that although the jolting of the train and carriage would give me a good deal of pain, it would not, in his opinion, retard my cure, it was decided to go up to Compañia at once. The *Wanderer* was to undergo a thorough overhaul of her rigging and hull, so we left her after five months and six days of life on board, most grateful to have been permitted to reach this stage on our journey so happily and so safely, and to have passed through the various contrasts of clime and seas and nations without let or hindrance, and to be safe again amidst the rustling trees and sweet scents of our old Chilian home.

## CHAPTER X.

### *AT HOME IN CHILI.*

FROM the 13th January, the date of our arrival at Coquimbo, until the 14th February, all on board the *Wanderer* were busily engaged overhauling the rigging, scraping the spars, and subjecting her to complete repair. I was glad to remain quietly amid familiar scenes and faces, until my ribs were mended, and all our party enjoyed the
15th Feb. rest on shore. On the 15th February I went down in the ship to Valparaiso, and spent a pleasant time amongst old friends, receiving a visit one day on board from Don Annibal Pinto, the president, whom I received with manned yards and all proper courtesies. The time was one of much national enthusiasm over the victories in Peru and the return of the victorious army. The election for the presidency which was also shortly coming off added to the general stir and excitement. The great division of parties in the country lies between the Conservatives, who have to carry on their backs the weight of the priesthood, and the Liberals, who, as in other countries, number in their ranks every shade of liberalism, and are only united when opposing the priestly power. I returned to Compañia on the 3rd, the party there having been well amused and occupied during my absence

RETURN OF COQUIMBO REGIMENT FROM PERU.

with riding, fishing, bathing, excursions to the port, a ball on board of H.M.S. *Triumph*, lawn tennis, and visits from the officers of the men-of-war visiting Coquimbo, amongst them our old friend the *Kingfisher*, and the *Turquoise*, *Thetis*, *Osprey* and *Shannon*. The capture of Callao and Lima had also kept the little town of Serena in a constant condition of fireworks and festivities. For myself there was, now that I could get about again, plenty to do in inspecting the works, and farms, thinning the plantations—which were grown beyond our wildest expectations, the blue gums in particular having averaged a foot a month since first they sprung above the ground—in visiting old friends, but above all, in making the most of the society of those dear ones from whom we should too soon have again to part, and winning the affections of our one little granddaughter.

Having some business to attend to in Santiago, I again went down to Valparaiso on the 26th March, arriving the **26th March.** following day at eight o'clock in the evening on a very dark night, and was greatly helped in picking up a good berth by the electric light from H.M.S. *Triumph*, which was lying in the roadstead.

On the 28th I went ashore, leaving orders for the ship to put out to sea if a "norther" should come on, as they often begin in this month, causing a heavy sea, which makes a ship labour and strain a great deal.

On the 29th I took the train for Santiago, being met at **29th March.** the station by two kind friends, to one of whose houses I was taken with that warm hospitality so general amongst the gentry of Chili, who seem as if they could not do enough to make one comfortable and happy in their homes. Two or three days were very pleasantly spent in paying and receiving visits from many old friends of my father's as well as my own, and being entertained with lavish hospitality. Before returning to Valparaiso I was anxious to have a look at an estate known as Culipran, of about 20,000 acres, belonging to Señor Don Ladislas Larrain, which lies some three leagues south of Melapilla, and fifteen from Santiago.

LAS CARDAS, CHILI.

We drove in a carriage with four horses abreast, and after leaving the town went at a rattling pace along a very good road, with large and fertile estates on either side. Relays of horses were provided at every three leagues, which stood waiting for us by the roadside.

On our way we had to cross, first the river Mapocho, near

CHORILLOS, AFTER THE WAR.

the village of San Francisco, and after passing Melapilla, the Maypu; the latter in a most unwieldy launch manned by four men, in which the carriage and our party were placed, whilst the horses swam. The Mapocho is a deep and rapid torrent, which runs into the Maypu, forming a river of considerable magnitude, quite impassable during the winter floods. It has trout, pejereys and lisas in it, but unfortunately for the sportsman these fish won't rise to a fly. Soon after crossing the river, we drove up to Culipran, after six hours and a half drive from Santiago—a good hour of which had been lost in crossing the Maypu in the clumsy ferry—and were soon after sitting down to an excellent Chilian dinner, washed down with good Limache wine.

During my stay I was busily engaged in visiting and inspecting the estate, with the intention of becoming its owner should it prove as attractive as I was led to expect from the descriptions I had received.

Friends in England would no doubt have said to me: If you want to buy an estate, why not buy one in England? To which I would answer, that in my eyes, estates in England are not so attractive, they are decidedly not so remunerative, neither is their tenure more secure, particularly after the experiences we have undergone regarding the respect shown to the rights of property in Ireland. The beauties of English scenery are great, but there is an air of artificiality about them; trees have been planted here, cut down there, a stream dammed up, a river turned, cuttings made, what is beautiful brought into prominence, what is ugly hidden from sight. Town life and English society show much the same features; large houses, fine streets, and the display of wealth hide from view the workhouse, prison, madhouse, and squalid poverty of thousands. Last, but not least, the climate is best described by quoting the Yankee, who said, "England ain't got *no climate*, but she's got *samples* of pretty near all sorts."

In Chili the beauties of the country are all natural ones, the varied forests, luxuriant fruits, and fertile fields owe little or nothing to the art of man, but spring in all their abundance direct from the lap of nature. The climate is a most beautiful one, bright and temperate; the soil rich and prolific, growing the vine, orange, peach, pear, apple, walnut and olive; and whilst in England you think yourself singularly lucky if your land pays you 3 per cent., in Chili the same investment will pay a good 10 to 15 per cent.

I found that Culipran was watered by a canal which takes its supply from the Maypu, and carries it about six leagues, and that the wild beauty of the country had not been exaggerated. The land where cultivation had been commenced was producing fine crops of wheat and barley, and the beautiful hill slopes supporting herds of thousands of cattle during five months in the year. Far away in the distance, in his mantle of eternal snow, stood out Aconcagua, looking down upon the gleaming Mapocho and Maypu rivers, as they rush along through the fertile plains of Santiago to pour their waters into the wide Pacific. My time, alas! was not long enough for me to see the whole extent of the estate, but I saw sufficient to determine me in becoming its possessor, and on my return to Santiago speedily concluded the terms of purchase.

12th April. I found on getting back to Valparaiso that the *Wanderer* had twice put out to sea to avoid northers, but that she

LA COMPAÑIA.

had come back again on the 11th, so that I was able to return in her to Coquimbo on the 12th, and found myself back again at Compañia on the 14th. It had been our intention to leave Chili about the middle of May, but finding our son Robert was on his way out from England

THE BRILLADOR MINE.

to join us, and that he hoped to arrive in Valparaiso by the end of May, it was decided to wait his arrival, the delay thus incurred giving me time to see a little more of my new purchase. On the 14th of May, after a quiet and happy stay at Compañia, my son-in-law and I started for Valparaiso, taking Pritchett with us, who was going to 14th May.

make some sketches in Juan Fernandez, as I decided that during our stay on shore the *Wanderer* should go down to this island, with the double object of keeping her out of the way of the northers, and of exercising the crew in evolutions under canvas. The 16th of May found us again at Santiago, under the hospitable roof of my old friend Errazuriz, and after a day or two's stay we went on to Culipran as on my previous journey, and spent five very pleasant days, chiefly in the saddle, inspecting the place, accompanied by Señor Gana, who was to manage the estate for the present. Suffice it to say that, on more intimate acquaintance, I was as pleased with the place as I had been at first, and that Black was as charmed as I was. We both left with regret to return to Santiago, and having said good-bye to our
27th May. friends there, got back to Valparaiso on the 27th to find that the *Wanderer* had looked in the same day, but put out again to avoid a norther which was blowing hard. Valparaiso, seen under the circumstances of this wind and a drenching rain, it must be owned, is not a cheerful place, but we had to make the best of it and wait as patiently as we could for the arrival of the *Magellan*, which was bringing
30th May. Bob. Directly after breakfast on the 30th we got the news of the *Magellan* having arrived at Talcahuano, so I sent off to the *Wanderer*—which had come in on the previous afternoon—to be ready for a start on the morrow, and we spent the day visiting friends, and winding up with a dinner at the club, at which we were joined by Bourchier, Lyon
31st May. and Böhl. On the 31st, at noon, we were on board the *Wanderer*, the mail steamer being reported nine miles to the southward, and soon after had the satisfaction of seeing her pass close by us with Bob on board. Our galley

LA SERENA AND COMPAÑIA.

was alongside almost as soon as she anchored, and in a few minutes we were welcoming our dear son on board the *Wanderer*.

After a hurried lunch we went ashore for an hour or two, picking up and bringing back Herbert Gibbs, who, we hoped, might have been able to come on with us to the Sandwich Islands, or further; this, however, he was prevented from doing, and only came back to say good-bye to our party

CROSSING THE RIVER, SERENA.

at Compañia, which we reached on the 1st June, to find everybody busily preparing for our departure, the time for which was now drawing near. In the midst of all our arrangements, I received a telegram from my friend Huneeus, in Santiago, begging me to bring Mr. Cuadros (a mutual friend, who represents Coquimbo in the Senate) to Valparaiso with as little delay as possible, as he was urgently required to prevent a vote of censure being passed on the Government. This telegram was delivered about 9 P.M. on the 6th;

1st June.

6th June.

and ordering a trap, I drove off at once into Serena to see if Cuadros was ready and willing to go, at the same time sending a messenger on to the ship to order everything to
7th June. be ready for a start next day. At ten o'clock next morning, Cuadros, with a nephew of his own, Herbert Gibbs and I, were alongside the *Wanderer*, to find that my letter by some mistake had only been delivered at eight the same morning, and that our engineers had the cylinder covers off and the boilers empty: everything, however, was being pushed on as rapidly as possible, and Mr. Boniface promised we should have steam up by eight o'clock that evening; so we spent a loafing sort of day, with occasional visits to the engine-room to see what progress was being made. At 7 P.M. Mr. Boniface had so well redeemed his promise that we were at full speed for Valparaiso, where we arrived, after
8th June. a most charming passage, at 3.30 P.M. on the 8th. The life-cutter was speedily lowered, and our friends with their luggage placed in her. At 4.30 P.M. she was back again, hoisted up, and the ship *en route* again for Coquimbo, where
9th June. we arrived at half-past two the following afternoon. I landed at once and drove off to Compañia, to resume my work and settle various business which had been interrupted by this political crisis.

14th June. On the 14th our farewell visits to friends and neighbours
15th June. were paid, and on the 15th we once more found ourselves settling down in our old places on board the *Wanderer*, having bade adieu with sad hearts to Compañia. Our daughter, her husband, and our little granddaughter came down with us to spend the last day and night with us on
16th June. board. On the morning of the 16th some young friends from Compañia joined us, bringing a good budget of English

LA SERENA, CHILI.

letters and newspapers, just arrived by the Panama mail, and we had a large and pleasant party—I cannot say a merry one—to lunch. At a quarter to four the parting words had to be said—no need to dwell here upon what is of itself so painful—and by a few minutes to four all were in their boats, the ladder was hauled up, the rigging manned, and three hearty farewell cheers given and returned as the *Wanderer* steamed off at full speed for Otahiti, and the little boats with their precious cargoes made for shore. Good-bye to the town of Serena, which we catch sight of from the sea. Good-bye to Coquimbo. Good-bye to the brown hills and woods of Compañia, to the dear relations and warm-hearted friends in the sunny Chilian land—and away again to the west!

LITTLE MAGGIE.

## CHAPTER XI.

*FATOU HIVA AND RESOLUTION BAY—TAOUATA.*

16th June. It was a peaceful lovely evening; the calm of all around seemed to suit our saddened spirits, and we steamed away over a glassy sea, without a ripple on its surface, watching the fast receding landscape and rugged hills, until night shut out all from our view.

17th June. The dead calm and the oily sea continued with us until noon, when a fresh breeze set in, so we feathered the screw,
22nd June. and continued under sail until the 22nd, when, owing to the wind failing us, we had to take to steam again. The weather had been very variable, the ship rolling a good deal from a high southerly swell, and as there seemed every probability of this continuing, it was decided to give up our visit to Easter and Pitcairn Islands, especially as it is impossible sometimes to communicate with the shore for a week at a time;
23rd June. and on the 23rd we accordingly shaped a more northerly course, which would bring us into the tropics, and under the
28th June. influence of steady trade winds. On the 28th we got into the tropics, and were rid of the southerly swell which had
14th July. tormented us so long; and from this date until the 14th of July continued under sail with an occasional resort to steam. (A full account of our proceedings will be found in the

FATU HIVA, MARQUESAS.

Appendix for those interested.) At 9 A.M. we sighted Fatou Hiva, one of the Marquesas group, and there was an eager clustering on the poop to see land again after our twenty-eight days from Coquimbo, during which time only two ships had been seen.

The outline of the island is steep and rugged, cut into with deep ravines. The more exposed parts are covered with sun-dried grass, but the hollows of the ravines were full of bright green orange and cocoa trees.

A bold precipitous cape, called Venus Point, rises from the sea to a height of 750 feet; rounding this, our anchorage opened out in a shallow bay, leading up to one of the large ravines. A little village stood on the beach, surrounded with palms; figures in red and white moved about; a bridge thrown across a cove foaming with surf, and a house with windows, showed signs of civilisation and provoked murmurs of disappointment from the boys. "I expect that's a hotel," said a voice in a tone of the deepest disgust, but presently a real canoe with a great log outrigger pushing through the surf to the quick beat of the paddles wielded by brown, tattooed, and almost naked figures, soothed the wounded feelings of expectation. There were two passengers in the canoe, one was a French officer of marines in temporary charge of the station, who spoke English very well, and who proved a most obliging cicerone. Then came more canoes, bringing cocoa-nuts, and then a general desire to get ashore. The landing was not an easy matter, and required a good deal of dexterity, but we were piloted by our new acquaintance, the French officer. The jetty is formed by a point of lava rock, jutting out far enough to reach the swell of the water just before it curled and burst in surf upon the shore, consisting of a steep bank of boulders, at the foot of a high black cliff. The galley was floated in on the top of the swell, stern on, until close to the rock, the rising water brought her to a level with the top of the jutting point, when as the boat was stationary for a second or two, as many as could scrambled out, until she was drawn in again by the next roller, when the rest followed suit. We were immediately surrounded by a crowd of laughing chattering natives, few of whom could talk English, fewer still French. Most of the men were naked, some sported a loose fluttering shirt, but to make up for deficiency of clothing they were most beautifully tattooed from head to foot. The women and elder girls wore long, bright-coloured cotton dresses, but no other garment. They had good features, and bright cheerful expressions, and fine agile figures, with dark olive skins. Some of the women had very pretty and delicately formed hands and feet.

A walk over the bamboo bridge brought us to the barrack, where there were about a dozen soldiers—a sufficient force with their breechloaders to overawe the natives now that nearly all their weapons have been taken from them, either by force or purchase. In front of the barrack some palms and other trees made a cool and shady spot, under which the native girls, in their orange, red, and white nightgowns gathered and watched us with a sort of mild affection. The following amusing little incident was also to be seen—an English man-servant, trying to keep up a respectable dignity, with a Marquesan beauty on each side of him holding his hands, their little brown fingers interlaced with his, whilst they gurgled and cooed in his face. Squatting near us was a fat old woman with yellow hair that stuck out horizontally like the branches of a cedar tree. She had fallen in with the new custom of putting on a nightgown, but she evidently thought it a foolish business, and the sleeves seemed to annoy her particularly, for whenever she wanted to use her hands or arms, she slipped them out at the bottom of her gown, raising the whole garment in folds to the necessary height to enable her to do what she wanted. Whenever we went ashore this old woman seemed to be lurking about on the look out for us, giggling and "carrying on" like a forward young minx of sixteen. Beatrice's long plait of hair attracted a great deal of attention from the native ladies, who followed her about, lifting it up and letting it fall on her back, and she kept very close to Bob, not at all fancying these familiarities.

We got some cocoa-nuts, oranges, and rose apples—this latter fruit is beautiful to look at, like a highly polished wax apple, so pink and white, but it is rather insipid, with soft white flesh, not unlike a lettuce, and a large seed in the centre, like a chestnut. We tumbled into the boat again with even more difficulty than we had tumbled out, and got on board to sit down to a motionless dinner table—a great pleasure after the constant rolling. It was a most perfect night, far into which we sat on deck, with senses soothed by the rushing of the surf upon the beach, and gazing at the fantastic spires of Point Venus looming grandly over our heads.

At 7 A.M. next morning our French friend took some of our party off in the dingey, to visit a spot of ground which is "*taboo*" and reserved as a place of burial. Very few natives were about, but we got some fresh cocoa-nut milk, one nut alone filling three tumblers full. 15th July

We then pushed our way through some hundred yards of cotton shrub, with its beautiful saffron flower, manioc plants with starry leaves, and sprouting cocoa-nuts, until we came to a steep and thickly wooded hill side. After a few minutes' climb we reached a terrace built on the side of the slope, with the basalt boulders from the beach. One stone in the centre of the face of the terrace measured over five feet high, and was roughly sculptured. On the platform were two coffins of hollowed cocoa-nut trees, slung between stakes, stuck up after the fashion of piled arms. Over these coffins were the remains of a kind of thatch of leaf matting. Skulls grinned at us from the hollows, between the roots of neighbouring trees. The coffins had fallen partly from their places, and skeletons, tappa cloth, strings of beads, and hair mixed with cocoa fibre, streamed over the sides on to the ground, presenting a gruesome spectacle. Bowls for holding food for the corpses lay about, a flask made from a gourd hung from a branch, and two roughly carved wooden gods surveyed the scene with a sort of drunken solemnity. The natives would not enter one of these groves on any account, and leave the bowls and gods to decay with the bodies. We then struggled further up the hill, through the thick wood, to visit other platforms, equally ghastly and horrible. In the passage from one of these charnel groves to another we passed one of those curious trees, the screw pine, whose roots branch out from the trunk two or three feet from the ground. Our guide told us that the Kanakas hold these trees in great reverence, calling them god trees, and never touching them if they can avoid it. It appears that after the Kanaka dead have been coffined and groved long enough for complete decay, some one is found of sufficient courage to cut off the skull and hide it, generally in the hollow of a tree. On our way back we found a regular nest of skulls in capital condition, and while Wetherall was examining one of them, a wriggling and scratching inside made him drop it like a hot coal, to see a lizard struggle out on to the ground.

After breakfasting on board we all started again for shore, the landing getting worse and worse, rollers breaking over the projecting point, keeping us waiting in the boat to catch a little one; all, however, got safe to land, with the exception of Captain Gordon, who missed the rock, and fell back between it and the boat, getting of course a complete drenching, so that he had to go back to the ship to change. The old girl with the yellow hair was still hovering about, but seemed rather shy of the ladies. Mrs. Lambert gave her a little mouth organ, at first she seemed rather doubtful about it, but at last she was induced to take it, and blow into it. When she heard the sounds it made she burst into a laugh, and covered up her face with her dress, raising it as described before as if she had committed some great breach of etiquette, and made off into the bush. After a little while she came back with two or three yards of sugar cane, with which she gently touched Mrs. Lambert on the arm, intimating this offering was in exchange for the musical instrument.

We strolled along a shady road, through a forest of bread fruit, cocoa-nut palms, guavas, oranges and limes, followed by a wondering crowd, who seemed to take the deepest interest in all we did. The chief was not at home, but we went into his house, built of bamboo, with a neat matting on the floor, a bench, frame for a bed, and a box which comprised the whole of the furniture. Then on again through the wood, until we came to a fallen tree, on which we sat, the crowds of followers all sitting in a row before us, watching every movement, and jabbering in their soft cooing language. Nearly all the women were decorated with flowers; one we noticed wearing white gardenias in her ears. They got up one of their native dances for our amusement, the men standing opposite to the women, singing a monotonous chant, and in the intervals between the verses taking hands and swaying about. The performance lacked spirit, however, partly from the fact of there having been a feast the day before we arrived, at which these dances had been kept up with great vigour all day. Our French friend told us that the natives take the most extraordinary pleasure in these dances. Near the hut of the yellow-haired squaw there was a square platform of boulders, which we were told was tabooed, and where no woman would go, as it had been the scene of human sacrifices, to which women were not admitted. One old man specially attracted our attention from his frequent loud and brisk laughs, and the impressive way in which he shouted "Kaoughah," a word which seems to do duty for every form of salutation. His beard of a sandy colour was carefully tied up in a knob (after death these beards are shaven off and kept as heirlooms); this particular old gentleman was a notorious cannibal. Another old fellow showed us one of his hands which was withered and distorted from a ball, fired by people whom he tried to describe to us by pointing down his throat, and working his jaws with ravenous snaps. It was very curious to see how utterly out of control the feelings of these people were. This last old man wriggled and quivered all over at the sight of a glass of sherry, and the face of a man we were buying some things from became distorted with rage at the sight of some nickel Brazilian money, which was among other coins shown him to choose from. It seemed that a Yankee whaler had come in a month or two before us, and done the natives on a large scale by giving them these coins as silver.

The main produce of the islands consists of dried cocoa-nuts, called copra, edible fungus, and rough cotton, which grows very luxuriantly and yields it is said three to four crops a year. In such a climate, however, with food in abundance, and fruits growing in wanton luxuriance, waiting to be picked and eaten, work seems to have no object; consequently not the hundredth part of what these islands might produce is grown, the little cultivation that there is, being the result of the uncertain efforts of a few runaway European or Yankee sailors, who, mated to native women, and free from trouble or care, live to a great age in these lovely islands where disease of any kind seems almost unknown. The general impression that we got of the natives was a pleasant one. Seeing them sitting hand in hand, clasping arms in chattering laughing groups, on the look out for our landing, we felt as if we were going to play with a pack of children, whose gentle gaze seemed to tell of no anxiety, or of any thought to give intensity to their character. Their life seems as near an approach to communism as it is possible to conceive, the communistic principle being apparently carried into their marriage customs. If one of them was smoking, you only had to wait for a few minutes to see the pipe or cigarette passed round the whole group. The French have put a veto upon human feasts, but the place and people are full of the memories of cannibalism. We got back to the ship for lunch, after which Captain Gordon, Bob and I, started in the galley under canvas to visit a bay about four miles to the north of Bon Repos, which was reported as well worthy of a visit, and where there was a Jesuit mission established, a Protestant one being established at Bon Repos under the direction of an English speaking Kanaka from the Sandwich Islands. The wind was baffling and uncertain, and it gave us some trouble to get to our destination, as it rushed down the mountain gorges that surround and back the little bay with such force that we could not get up under canvas, and it was all that six stout oarsmen could do to pull us up to the head of this wild but beautiful little creek. It was encircled with perpendicular cliffs and pinnacles in all directions except one, where there was a bit of boulder beach, some seventy yards in length; at the back of this stood a little spired church, with school and mission-houses and several native huts. The landing place was on a narrow ledge of rock, with an overhanging perpendicular cliff above it. The back lash from the beach, together with the incoming rollers, made landing both difficult and dangerous. Amongst the men on the beach who had come down to gaze at us was a native of Manilla, who spoke English very fairly; he came alongside us in a canoe, and begged us to land, saying that the priest and school-children were in the chapel but would soon be out; in addition, however, to being anxious to save the daylight, I had no fancy for risking the galley on the rocks, so we turned towards the ship again, passing on our return several curious blowholes, that threw up immense jets of water in the air with a noise like distant artillery, as the waves drove in beneath them. Coasting along on our way back the scenery was most beautiful and romantic, little glens ran up from the shore, clothed with cocoa-nut palms, each having its native hut, and at the back stood up the rugged peaks of the mountain range, whilst the setting sun fell upon the flashing spouts of water that rushed from the blow-holes, and lit up the whole scene with a blaze of colour. The scene was exactly what a vivid imagination might picture as a dream of the South Pacific Islands. In the evening Mr. Gellineau dined with us and gave us a good deal of information about these islands. He spoke English well, having lived for some time in London, and confirmed our opinion of the natives, saying they are most docile and inoffensive, and very fond of Europeans, but that after drinking the fermented liquor of the cocoa-nut called "puy" they seem to be capable of any atrocity, whereas the "kava" only seems to produce drowsiness. The French punish any excesses very severely, transporting the delinquents to another island for from three to six months, a punishment they particularly dislike. They seem to have no religion, although gods in stone and wood abound, but they appear to have no special respect or veneration for them, except when they take them out on a fishing or warlike expedition, when if successful, the god is made a great deal of, offerings of food being placed by its side, which the priests take good care (like those of Baal) to eat up shortly after; but if the expedition turns out a failure the god is beaten to pieces with clubs and stones. Cannibalism has practically ceased, since the islanders have had their arms taken from them, for as there are no wars, there are no wounded or prisoners to furnish food for the cannibal orgies. Several hardened old sinners, however, were pointed out to us, who belonged to some of the best old cannibal families, who smiled as sweetly and looked as innocent, as "the bage unborn."

The supply of provisions is bad, for although there is an abundance of fruit, there are no vegetables, and very few fowls or eggs. The fish that we caught also were indifferent, the best being a sort of brill. At the Jesuit Mission there is a flock of sheep, but we heard of them too late to get any, and the pigs are so thin and small that they have to be fed on board at least a fortnight before they are worth eating. The French garrison at Fatou Hiva consists of a sub-lieutenant, with a sergeant, corporal, and ten rank and file. Some of these, however, were away when we were at the island, having gone to Hiva-o-a, where the anniversary of the destruction of the Bastille was being celebrated.

July. We left Bon Repos at 6 A.M. for Taouata, passing the blow-holes again, and watching the various beauties of the coast, as we steamed by Motau and through the Borderlais Straits, between Hiva-o-a and Taouata, and anchored in Resolution Bay at 3.50 P.M. after a run of fifty-five miles. The general aspect of this island was not so grand as Fatou Hiva, but more suggestive of sheep-runs. It was here that the natives a year or two ago repulsed the crew of a French gun-boat, and cut a Swede in pieces, but were brought to reason by the French admiral, with a mixed brigade of 800 men. The harbour of Resolution Bay is formed by one of the central gullies, dropping its lower end into the sea. The general lie of the land, and the rocks—a confused medley of volcanic ash, basalt, and porous pumice-like substances—left little doubt that we were at anchor in the bottom of an old crater. The solitary gendarme in charge of the island, came alongside in a good whale-boat with tricolor flying, and acted as our guide ashore. The landing was somewhat better than at Fatou Hiva, but the island itself is certainly not so interesting. The natives showed signs of more civilisation, they were dressed in shirts and trousers, and some wore a French kepi, but they lacked the simple friendly air of the people of Fatou Hiva. An American carpenter, who had set up a bench, on being interviewed, described the island as dull; beyond this he seemed to have no impression upon his mind. "Curiosities?" he didn't know of any, but would send some of the niggers to look for some. Thereupon he spoke to those who were standing by, in some unknown tongue in a loud commanding voice, to which they paid not the slightest attention; experience no doubt had taught them neither to fear his wrath nor relish his dealings. Presently the doorway of a hut hard by was filled by a large square-mouthed native, who called out in English "Send that man to speak to me." This was the king of the place, who had been employed by Captain Medlycot of the *Turquoise* to pilot him about the Marquesas. "How many stripes your captain?" said he. When told, "three," he replied with an air of great self-importance, "My friend captain have four," evidently looking upon us all as very small potatoes. Our gendarme showed us two large stone block-houses, with two stories, the lower one loopholed for musketry, and the ruins of a ten-gun battery, the solid nature of which, combined with the firm roads in the island, and the style of the landing place, seem to show that the French must have intended to make a formidable post here. There is a Jesuit mission also, in charge of an old curé, who has been settled here for a long time. He showed us his neat little chapel, and school girls dressed in long bright-coloured cotton gowns, but they were not so nice looking as the women at Fatou Hiva.

Coming off to the ship an amusing little incident took place. Harris had bargained with a ferocious looking native for a piece of the hard wood the island produced, and just as we had tumbled down the shelving rock into the boat, and were rising and falling in the surf, the wild man appeared with the wood. Harris gesticulated frantically for the wood, the native replied in the same style for the money; orders being given to Harris to look sharp and close the negotiations, he threw the money to the man, who spat at him in defiant derision, and made off with the wood, exulting that for once in his life anyhow he was even with the swindling white man. Unfortunately for him our friend the gendarme was looking on, and just before we weighed anchor, the bit of wood was brought off in a canoe. It was an evening to be remembered for real comfort, for as soon as we cleared Resolution Bay, we passed into a strong breeze blowing favourably over a still sea, the islands lying as a breakwater to windward. All fore and aft canvas was set, causing a down-draught of coolness that freshened the ship throughout, and sending us to sleep in a fashion to which we had been strangers almost since leaving Chili.

## CHAPTER XII.

*NOUCA HIVA.—PEACOCK ATOLL AND TAHITI.*

The brisk breeze drove us along at such a good pace that at daybreak we stood into the entrance of Comptroller Bay (having completed the seventy or eighty miles between the islands in about eight hours), a fine sheet of water, branching off into three nice harbours, by far the best we had seen in the group of islands, and surrounded by lofty mountains with smiling valleys. Running along the coast watching the sun rise, and the effects of the changing light on the green foliage, and outlines of the mountains, we entered Hakachou Bay, passing between two rounded islands, known as "The Sentinels" and steamed towards the line of rich groves, whose green, relieved by some white houses, a landing stage and a schooner, marked the position of the seat of the Marquesan Government, opposite which we anchored. A boat soon shot out from the side of the schooner, having on board the harbour master, who put several questions to us. "Who were we? Whence did we come? Were all on board in good health?" All of which being answered to his satisfaction, he was joined by two gentlemen, one a Mr. Hart, of the firm of Hart and Co., of Otahiti, who have branches in all the most important islands of the South Seas; he is English by birth, but a naturalised American, the other was Mr. Brown of the same firm. Mr. Hart promised to provide us with fresh bread, eggs, milk, and water cresses—the only vegetable grown in the island. A schooner runs monthly between San Francisco and Tahiti, taking Nouca Hiva on her way, carrying mails, and bringing potatoes and other vegetables, but she was not due until after we had left. Mr. Hart further promised to kill an ox and send us half, with some sheep and poultry for our voyage to Tahiti. Some of our party asked about curiosities, but they were pronounced scarce, "very scarce and dear." After these visitors had left, we had service, and sat lazily about, for the heat was intense. Later in the day, Mr. Brown returned, bringing with him M. Gazengel, the Commissaire de Marin for the Marquesas, a lively Frenchman, who spoke English very well; and in the afternoon we all went ashore. Mrs. Lambert and I, called on the Resident or governor, Lieutenant de Vaisseau Chastanie, a very agreeable gentleman with a pleasant wife. Having chatted for some little time, and asked them to come off to lunch next day, we took a stroll in the village, followed as at the other islands, by a crowd of smiling inquisitive men, women, and children, but we found it dull and tame after Fatou Hiva, as the natives were dressed, and only the elder ones were tattooed. There was a shop too, with all sorts of commodities for sale, dresses, boots, caps, hams, &c., and finally a Hotel of all Nations; so we wended our way back to the landing stage, and off to the ship by dusk.

18th July. Bob and Wetherall were up early in the morning, having been told that there was something to shoot in the shape of wild goats, and were ashore before seven. Thinking it best,

AHI ATTOL

however, to make quite sure about the game laws in Nouca Hiva, they called at the Resident's house, to ask his leave. He had evidently jumped out of bed to receive them, but was most obliging, "Kill" said he, "as many wild goats as you like, and I hope you may enjoy your climb after them." (There was something rather sinister about this remark.) "Take a glass, look about you, when you see one, climb about the rocks and precipices, till you get within shot. I *not* shoot them." (Sensible man, he knew better.) Armed with this authority, away went the two sportsmen. The road which led to the governor's house, ended with his garden and orchard, on the far side of this they found themselves at once in a thick jungle of thorny dwarf acacias. Pushing their way through, they were soon rewarded by hearing a rustling noise near them. Rifles tightly grasped with intense excitement they waited. It comes nearer, what is it? Hush! Where is it? Look out! Here he comes! Then exclamations of disgust, as a mild eyed *cow* emerges from the scrub. Better luck next time. On they go, more rustling, more excitement, more disgust, more cows! At length there came a quicker tread, and the passing of some thing at a more rapid pace going up towards the hills, and on the bare side of a ridge above, a herd of six wild goats was seen climbing on to the top, and dropping one by one out of sight on the other side. A snap-shot at the last one, not a hundred yards off, was rewarded by the welcome thud of the bullet, added to which an upward leap of the goat filled the mind with satisfaction. On reaching the place, however, nothing could be seen of the animal; but a track of blood led along the other side of a knife-like ridge which made a regular rim to a volcanic basin, and on the other side of a jutting rock the goat was seen;

after three or four more shots, he tumbled down the face of a steep rock. The face of the rock was broken with lumpy tufts of long grass, so that Wetherall was able to let himself down to a ledge some forty feet below where the goat lay. And now, had there been time, and if any one had been inclined to listen, was a great opportunity to draw a fine moral lesson upon the vanity of gratified desires. There lay the goat, but what was to be done with him? It was impossible to carry him up the rock again, for the grass tufts came out in an unpleasant way with Wetherall's weight alone. No, there *was* nothing to be done but leave him there, crawl up again, and get back to the ship as quick as possible, hurried on by a most earnest longing for breakfast and something to drink.

At noon the Resident and his wife came off and saw all over the ship, and told us that he had ordered six ponies to be ready at two in the afternoon, two of them with side saddles; so after lunch, having landed the Resident, Nellie and Beatrice, escorted by Dr. Gray, Pritchett, and the two boys, started for a ride. From the ship we could see them wending their way up a steep hill-side, from the top of which they had been promised a view of three beautiful cascades; but to the top they never got, for the way was so steep and rocky, that in places they had to get off and let the ponies scramble over the obstacles, and follow on foot; in addition to which the ponies were small and feeble, and the time short, as they were charged to be back by sunset at latest. In spite of all the difficulties, however, they thoroughly enjoyed the ride, through groves of palms, plantations of cotton and sugar cane; and as they mounted higher, through guavas, limes, and oranges, from which they could pick the fruit as they rode along. The highest peak on this island is 3,840 feet, and the inhabitants number about 500. The French staff consists of the Resident, who is the head official over this, and all the other islands of the group, but is under the orders of the Governor-in-chief of the French possessions who resides at Tahiti, and makes yearly inspections in a man-of-war. The present Resident introduced the ponies about a year ago from Tahiti, to the great delight of the native chiefs who are anxious to import some on their own account. Sheep, goats, pigs, fowls, and cattle, are wild all over the island, and with the exception of the horned cattle may be shot by any one who cares to do so. It is curious to hear in the morning the wild cocks crowing from the hills around. Captain Cook introduced the pigs and fowls, but no one seems to know from whence or when the cattle and sheep came—very likely at some time from Mexico. There is no indigenous animal of any kind in the island, excepting the sea birds and some small doves, also most likely imported in some chance way, and there is only one other French official besides the Resident, who is harbour master and pilot. The traders are English, American, or German.

In the afternoon I went ashore with Captain Gordon to get tracings of a new French survey of the Low Archipelago, which differs very considerably from the Admiralty chart we had, and after finishing our work, drank a farewell glass of sangaree with M. Chastanie, and looked over his fine collection of native weapons, dresses, and ornaments. He most generously presented us with several of the old men's yellow beards, which are objects of great care as their owners feel old age stealing on them, and we were told that a very fine flowing sample would bring fifty dollars. Before we left, 19th July.

Mr. Brown came off with a very good collection of native implements and other matters belonging to the Roman Catholic bishop of the isles which he was willing to part with, and which I bought to add to the few things I had already collected. He also brought with him a very handsome native to put some of the things on, and give us an idea of a Marquesan on a gala-day. Presently he appeared, looking splendid with a circlet of waving feathers round his head, tufts of beard, tappa cloth, and hair round his loins, a leaf fan in one hand, and a wand ornamented with hair in the other, and his body most beautifully and delicately tattooed.

All strangers at last having left the ship, we steamed off and along the southern side of the island, close in land, passing numbers of beautiful little bays and inlets, with palms, woody hill-sides, and peaks. At Chichakoff Point we struck across to Hacaeton Bay, the northern extremity of Roa Pua, or Adam's Isle, which we reached about two in the afternoon. It is the grandest collection of rocky towers and cloud-capped peaks in all the Marquesas. We stood into the entrance; but a line of green and broken water across the mouth of the harbour decided us not to try the passage, so we kept close along the south end as far as Obelisk Island, passing many lovely spots, but nothing that looked inviting as a harbour. Roa Pua, although a small island, is the most thickly populated of all, and the natives have given the French a good deal of trouble. Admiral de Petit Thouars found it necessary to take the *Victorieuse*, and two gun boats with 500 men to avenge an outrage committed on some sailors who had come ashore from a gun-boat. The natives on the arrival of this force gave up the guilty ones, and surrendered their arms, but the French still keep fifty men

LOST AND SAVED

of the Infanterie de Marin there, the largest, indeed the only garrison in the islands. Before 5 P.M. after a most delightful day, we stopped steaming, and setting all plain sail, to a pleasant S.E. trade, shaped our course for Peacock Atoll in the Low Archipelago. As the sun went flaming down, the last of the Marquesas was dimly visible in our wake, and we all agreed that our voyage had given us nothing more attractive than the days spent in these beautiful islands.

After two days' very pleasant weather, we sighted Peacock 20th, 21st July.
Atoll at 3 P.M. on the 22nd. "Sail on the port bow," 22nd July.
had been first sung out, but after was corrected to "Palm-trees on port bow," and soon we seemed to be looking at trees growing out of the sea, until as we got nearer we could distinguish the white coral bases. At six o'clock we were close to the entrance, but, unluckily, there was not enough water for a vessel of our draught, so taking in all canvas, we got up steam, and lay off and on until next morning. At day- 23rd July.
light we were still within sight of the Atoll, and as we came up to it could see the wreck of a schooner lying on the southern side of the reef. The night had looked somewhat threatening, and we had been afraid that we should not be able to land, but our fears were groundless, for after breakfast, about nine o'clock, all the gentlemen, Nellie and Beatrice, and the two boys put off in the galley, under a blazing sun, and with an oily calm upon the sea, which was most wonderfully clear—the white coral throwing up the light from the depths below. At every stroke of the oars the water seemed to grow paler and of a more greenish blue, and when about a hundred yards from the gleaming white shore, we could see beneath us the coral groves with their fantastic branches, and shadowy grottoes, out of which came, darting across the

which marked the residence of the English consul. Looking towards the shore you might imagine yourself to be in a river or lake. There was no sloping beach or shingle, but a bank of green turf descended to the edge of the tranquil water. Further back ran a road, along which moved carriages and horses, and bright-coloured figures. To the left stood houses enveloped in woods, and a broad grass quay, against which a schooner and some large trading-vessels were moored, whilst a little further out lay two French ships of war. Between the quay and the outer reef was a small island, wooded with cocoa-nut trees, which is sometimes used for quarantine purposes. As it was Sunday we passed a quiet evening on board, enjoying the sunset over the dark blue ocean, relieved by the white breakers on the reef which encircles the island
25th July. and forms the harbour. We were up at dawn to visit the market, which opens at 4 A.M. and closes at 6 A.M., and found a string of Chinamen on each side of the road, sitting by lamp-light with their wares in front of them upon the ground—nice-looking radishes, carrots, lettuces, cabbages, prawns, crayfish, and a variety of many-coloured fish. Of all these good things we laid in a good stock for breakfast, which was most thoroughly enjoyed. After lunch we went to call upon the consul, Mr. Miller, and his wife, a very charming lady from Lima, who was pleased to find we could talk Spanish; on Mr. Darsie, and others for whom we had letters, and also to visit Major Hill's ice manufactory. The young people had the dingey out and amused themselves with paddling about and visiting the reef, with its wonders of curiously coloured fish, black, canary-coloured, and white with blue heads and tails. Mr. and Mrs. Miller came to
26th July. lunch, and in the evening we all went off to the reef again

to see the natives fishing by torch-light, a very pretty sight, the fishermen a most happy, laughing set of fellows. It was bright moonlight when we returned, to find our men gathered round the dripping figure of a Chinaman, who had swam off from shore, having escaped from prison, where he was confined for breaking into and robbing some house in the town (he described this act as "holding articles for a friend") and begged to be allowed to remain on board as a stowaway. We could not listen to this, however, and sent him off to one of the French men-of-war, from whence he was sent back to prison.

This was a very pleasant, easy day, Mrs. Lambert, Bob, 27th July. and I went away in the galley for a sail, and in the evening the ladies went to see some native dresses being made that they had ordered, and then with Mrs. Miller to the queen's palace—much more like a cottage than a palace, with a veranda and an entrance court with grass and trees. The king has a more pretentious-looking house on the other side of the road, for this royal couple do not live together. The king has made over, for an annual payment from the French Government, his right of sovereignty, and his heir will be only a Tahitian gentleman. When we reached the veranda an Albino woman-servant appeared and showed us into a simple room with a few pictures; and in came her majesty, a tall, fine woman of about twenty-five, dressed in a loose white gown with a plaid bow at her neck, her black hair hanging down her back. She has a very pleasing face, and speaks English well, having been educated in Sydney. Mrs. Miller presented us, and we sat down and chatted, and invited her to lunch with us on board the yacht, which she accepted. We then returned on board, having ordered two carriages to be ready next morning at 5 A.M. to take our party (with the exception of Mrs. Lambert and Miss Power, who preferred remaining quietly on board) for a two days' drive over the island.

28th July. Accordingly, next morning we were ashore by dawn, the air charmingly cool, the bright stars, the streaks of orange in the sky deepening in colour every moment, the twinkling lights on the road, the gloom of the avenues with dim figures of natives and Chinamen going to market—some with long poles on their shoulders, the outlines of the palms, the great banana leaves—all giving a fairy-like and magical atmosphere to our exit from Papeete. There were nine of us besides the drivers packed into two curious conveyances, something like mail phætons, except that we had three rows of seats, instead of two. At 10 A.M. we reached Paea, where at a small hotel kept by an old soldier, M. Chamine, we got a most delicious breakfast, quite as good as anything that a Paris restaurant could turn out. We sat in the open air under a veranda close to the pale green sea, with its fringe of white foam, and looked out to the deeper blue of the ocean beyond, and the beautiful mountain peaks of the island of Morea in the far distance. Around us the luxuriant foliage of the palms and many other lovely trees, kept the air cool and fresh, and the white coral beach sparkled in the sunlight. The drive was a beautiful one, but it was sad to see what were once thriving estates, now lying waste, weed-grown and deserted. After breakfast, an excursion in canoes was made to the reef, and after that a bathe, and then on in the carriages again through scenes which have been well described by other and more accomplished writers; by vanilla plantations, sugar and cotton plantations, alas, sadly neglected and abandoned, and through the estate managed by Mr. Stewart, who had at one time, 1,500 Chinese coolies upon it, their dwellings, and the manager's house—all now in ruins. We arrived at Papeuriri about 5 P.M., and sat down to an excellent dinner, finding the hotel kept by another old retired soldier with his wife, a very pleasant woman who made us most comfortable, but as they could not take us all in, some of our party went off to the house of the chief of the district.

## CHAPTER XIII.

### *MOREA AND BORABORA.*

29th July. NEXT morning at five o'clock we made an early breakfast, before continuing our drive. The scenery on this second day was of a different character, for our road for a great portion of the way ran through a dense forest, and over muddy bad roads. Sometimes we left the sea-coast and ascended the hill-side, and then came back again to the foot of precipices that towered above us, clad with ferns, and plants with curious, giant leaves. At 11 A.M. got to Taravao where there is a barrack and fort occupied by a few soldiers, and here we sat down in the little inn—kept again by an old soldier—to a most delicious breakfast of prawns, crayfish, oysters, and other delicacies, that no one would dream of finding in such a spot. We were to spend the day here, so strolled about, and bathed in the warm sea. In the evening the Commandant of the out-post joined us at dinner, after which he took the gentlemen of the party to his quarters to sleep, the ladies remaining at the hotel.

30th, 31st July. The two following days were spent on the return journey; we had seen on this drive a good deal of the island and its varied beauties, but the general decay that seems to be going on unchecked throughout this beautiful country is most mournful. Roads that were once good, fast becoming impassable from want of repair, fine estates with buildings mouldering away, and produce rotting on the ground; the natives gradually dying out, and a sort of hopeless, helpless indifference displayed on all sides. The French inhabitants themselves denounce the wretched misgovernment of the country; alas, for the folly and indifference of the English who have allowed these lovely possessions to slip out of their grasp!

On our return to the ship, we found that during our absence she had been fresh-painted and the rigging all set up, so that she looked as smart as one could wish. The ladies who had stayed on board, had been busily occupied visiting the sights of Papeete, amongst them the tomb of the great queen Pomare—a curious structure, made of coral, and decorated with ornaments made from the red volcanic earth. It stands on a peninsula, stretching out into the sea, shaded with a variety of beautiful trees. They called also on the Roman Catholic bishop, who showed them his garden and beehives, and found he had been in Chili and spoke Spanish. Another day they drove with Mr. and Mrs. Miller, and dined with them in the evening—the dinner cooked by a Chinaman, and they were waited on by a Tahitian servant in white trousers, with his shirt hanging outside, and bare feet. The shops of course had been visited, they are mostly kept by French people, and on the Sunday they had gone to the native church, the service being conducted in the Kanaka language. They found the building of wood, cool and large, filled with the people in their picturesque dresses of bright colours, and straw hats with gay ribbons and flowers. The church smelled rather strongly of cocoa-nut oil, as the people use 31st July.

it freely on their hair and bodies. The missionary gave out a hymn; the people sitting struck up, one or two at a time, then others joined in a curious melody, with each verse ending in a sort of grunt, but very good time was kept. Then came a prayer, and then another hymn to the tune of "Just before the battle, mother," this was very prettily sung in parts, the men keeping up a sort of bass drone.

1st Aug. At 3 P.M. to-day the galley was manned and sent to meet the queen, who came off, bringing her brother and sister with her. He is a very fine-looking man; and all the party spoke English well. The queen was dressed in a loose dress of brown brocade, trimmed with yellow satin, and a straw hat; the sister in yellow cambric. In our visitors' book the queen's name stands as Marau Pomare. We found her very pleasant and chatty, and she told us some interesting stories about the little child-queen of Borabora whose kingdom we hoped soon to visit.

2nd Aug. On the following day we received a visit from the king; he can only speak a few words of either English or French. With him end the honours of the royal house of Pomare, as he has sold his right to the crown to the French, for an annual payment of 60,000 francs. The Roman Catholic bishop also paid us a visit; a very intelligent pleasant man, the founder of the various Jesuit missions in the islands, both himself and his staff of priests are highly and justly respected by Protestants as well as Catholics.

5th Aug. On the 5th we had a pleasant farewell dinner at M. Chessi's house; it was the anniversary of our departure from
6th Aug. Cowes, and next day we left Papeete Bay at 11 A.M., and steaming through the narrow passage, directed our course

towards Morea, and entering the opening in the reef, anchored in Talis Bay at 1.45. The harbour here is long and narrow, and singularly beautiful, with its woody hill-sides, and flowering trees, descending to the water's edge. Far above, the rocky peaks look as if they had been cut with some giant's scissors, they stand up against the sky in such sharp and fantastic shapes—but the scenery has been well described in *South Sea Bubbles*. The natives here seem to suffer much from elephantiasis, and we found the mosquitoes particularly venomous and vivacious. The French flag flies here, as it does at Tahiti; and as soon as we anchored we were boarded by a very polite lieutenant of marines, who offered us his services.

In the cool of the day we went ashore to call on Mr. and Mrs. Godeffroy, to whom we had letters. The firm which he belongs to owns the only cotton and sugar estate of any size in the island, which is worked by Chinese labourers. These fellows were just commencing the annual feast to their departed friends; and when we got on board again we watched them arranging little bowls along the beach, which they filled with tea and fruits, after which lights were placed in front of the bowls; later on their huts were illuminated and decorated with paper lanterns, joss-sticks, and strips of paper having the names of their gods written upon them in gilded letters. From this island thousands of oranges used to be exported to San Francisco, but the enterprise did not pay, and the fruit, which is very large, now falls and rots on the ground for the pigs to eat. Our men were told to carry off as many as they liked to take.

We passed a quiet Sunday morning as usual. Dr. Michaeli, an Italian gentleman, long resident in the island, joining us at service. 7th Aug.

8th Aug. Shortly after breakfast on the Monday morning, Mrs. Lambert and the two girls went for a sail in the galley to the reef, for these reefs are a never-failing source of wonder and delight, with their branching coral, and wondrous fishes seen beneath in the transparent water; but I am afraid they have been so often alluded to that I will not weary my readers with any further description of them. We bought some fish of a native returning from the reef with his supply for breakfast, one was as brilliant as a parrot, with all shades of green, red spots and stripes, and purple fins. The man assured us he was good to eat, but we rather doubted his extreme beauty of exterior. Afterwards we strolled about the village, surrounded by the men, women, and children of the place, hailing us with their cheery "Ya Rana." We sat down in one of their huts, whilst boys climbed up the trees to throw down cocoa-nuts, the milk being most refreshing and excellent. These simple people seemed anxious to give us anything and everything they had, without the slightest anxiety for a *quid pro quo*. They are fine and tall, and very good-tempered and pleasant. We saw the remains of what was once a Wesleyan chapel, built of coral, but the roof was blown off some time since, and as there are no Europeans now to repair it, the building is fast becoming a ruin. The French resident came off to lunch with us, bringing a present of some island honey, the bees taking their supplies exclusively from the orange and citron blossoms. Bob and Wetherall went out shooting, and were more successful this time, shooting a duck, and a wild cock with plumage as gay as a pheasant. They also saw wild pig, and got one with the help of a dog they took with them. At four in the afternoon we left the little bay, and

RAIATEA REEF

after a beautiful night, arrived next morning at the island of Raiatea.

The entrance to the harbour of Utuma is remarkably narrow, but clearly marked by a palm-clad island on each side. Mr. Keane, the representative of the Companie Commerciale de l'Oceanie on the island, promised us supplies of beef, bread, and milk, and we settled with him to make a tour round the island in the steam cutter, towing the galley. After a walk through the straggling village we came back to the ship, and then as usual went off to the reef, and on our return, canoes came clustering round us, with natives offering shells and fruits for sale, which they were only too glad to exchange for the little mouth organs which we had supplied ourselves with, and which seemed to give the greatest pleasure. 9th Aug.

After a scrambling breakfast next morning, our little expedition left the ship's side, and touching at Mr. Keane's mole, picked him up, and steamed away inside the reef for about two hours, when we stopped to see some sacrificial altars, made of two rows of large stones set on end, the gaps being filled in with smaller stones, beneath which appeared the remains of human skulls and bones. On the path which led to this place of sacrifice lay a large stone with two hollows scooped out of it. Upon this stone they say the victims were killed, cut up, and divided amongst the witnesses of the sacrifice, the bones being afterwards thrown upon the altar. The natives, who clustered round us, seemed to take no interest in our proceedings, or show the slightest dislike to our grubbing about amongst the bones of their ancestors. We then visited a monument to the first king of Raiatea, a man said to be of gigantic proportions. A 10th Aug.

drenching shower passing over, we gladly sought shelter in a hut, taking to the boats as soon as possible to escape the mosquitoes, which had persecuted us in the orange groves most abominably. As we got round to the south and west side of the island, the scenery became much wilder than in the harbour at Utuma, and by four o'clock we had completed the circuit of the island.

11th Aug. To-day the whole of our party went to pay their respects to the queen. She is a very tall woman, not less than six feet high, and quite devoid of the pleasant polished style and manner of the Queen of Tahiti. The palace is a large bamboo hut, one half of which has a raised boarded floor, the other half being on the ground. A portion of the upper part was screened off, with a cotton curtain, forming her majesty's sleeping apartment. We sat in a row on chairs, and with Mr. Keane's help, who kindly accompanied us, carried on a little conversation. On our expressing a wish to see the ex-king, he came in, a very tall man, at least six feet six inches, and not unlike Abe Lincoln. He spoke a little English. We invited him and the queen to come and see us on board, which they consented to do, so we went off to send the galley for them. They stayed with us for nearly two hours, during which time the old king drank about two-thirds of a bottle of brandy, and when he was going away asked for a bottle in paper to take with him, which, having got, his royal highness went off happy and rejoicing.

The history of this old king and his daughter seems to be, that the French induced the old man to accept the protection of France, and to hoist a protectorate flag. No sooner, however, had the French man-of-war left the harbour than his people rose against him, dethroned him, hoisted

BORRA BORRA.

their own flag, and placed his daughter on the throne. The bribe, however, of a few hundred dollars, was too much for the queen's patriotism, and the protectorate flag was again put up, and still flies, in utter defiance of a convention existing between France and England, by which the former was allowed to occupy Tahiti, the independence of all the remaining Society Islands being reserved. In the evening Mr. Keane joined us at dinner, and our pleasant little stay at Raiatea came to an end.

Turning the ship round we steamed off at 7.30 A.M., and keeping inside the reef, coasted along the south-west shore of Tahaa Island, and making our way through the western reef, shaped our course for Olea Vanna Narrows, in Borabora, where we anchored a little after noon. We found a very pretty and picturesque harbour, and, strange to say, the place not yet occupied by the French. The pilot whom we had to take on board outside the very narrow entrance to the harbour, told us that on anchoring, and before communicating with the shore, it was customary to announce the ship's arrival to the queen. Mr. Tyacke was accordingly sent ashore for this purpose, while some of our party followed in the galley. On arrival we found a very good stone mole, the best seen as yet at any of the Pacific Islands. Mr. Tyacke, having discharged his duty, met us with the intelligence that the little queen (of whom we had heard before) had received his announcement through an interpreter, and said that our ship was welcome, and that she would be pleased to see us. Upon this we went on to pay our respects to the little sovereign! A bright sprightly little creature of nine and a half years old, who received us very graciously, evidently taking a great fancy to Beatrice, whom

12th Aug.

she invited to stay and spend the afternoon, but as it was rather late, we promised she should come back the next day. Moe, the little queen's mother, was also present, as well as her uncle, who acts as prime minister, and practically governs the island in his niece's name. Moe is a very pleasant person, and speaks English well. Strolling through the village after this visit we were much struck by the well-to-do happy look of the natives, and by their size. We called on Mr. Blackett, an old resident of this, and the adjacent atoll of Tubai, where he had been for many years engaged in the pearl fishery. It is at Tubai that the bos'n or tropic birds are said to be so plentiful, and as we wanted to get some of their beautiful red tail feathers, Mr. Blackett promised to pilot us himself if the wind should be favourable for a landing on Monday. Coming back to the ship we paid a visit to a little French cruiser, and then ran alongside an Auckland schooner, to see if we could get any information as to how we should see some diving for pearls. The skipper, however, had a party of friends on board, and was not in a condition to give us much reliable information, so we left him alone, proposing to call next day.

13th Aug. At ten o'clock we sent off for the queen, and she came in state with her mother, another lady, and six chiefs. The queen was dressed in yellow silk, gold necklace, a hat with a pink ribbon, and wore high-heeled French shoes which evidently inconvenienced her very much—not to mention her garters, an evident source of worry; at one time during her visit the prime minister was seen trying to arrange these more comfortably for her. Moe was dressed in black silk, and the third lady in pale yellow grenadine with white embroidered petticoats. We had heard a good deal of this juvenile queen, and of her spirit in resisting the French. She ascended the throne at three years old, being adopted by her uncle, who was husband to the late queen. He is an elderly man, tall and slight, to whom she constantly referred. The queen fidgeted about all over the ship, and taking Beatrice by the hand made her show her everything. She was delighted with the piano, kicking her feet about in time as it was played. Before leaving she asked the girls to go and see her in the afternoon, which they promised to do, so after lunch Beatrice and Miss Power went off to return the visit. They found she had taken off her grand dress and put on a loose native one, and later on when we called for Beatrice in the galley, we saw the queen, having got rid of her visitors, racing about with bare feet in comfort. The next day was a Sunday, and from the deck we saw the people flocking into chapel at 7 A.M. and again at ten, for Sunday is kept here very strictly, and the people very orderly and quiet.

At 8 P.M. every night a bell is rung to call all home, and at nine it rings again, after which hour no one is allowed to be out. It was a damp muggy morning, but cleared sufficiently for us to have service, to which the queen, attended by her minister and six chiefs, came, also Mr. Blackett. The chiefs all looked remarkably uncomfortable, three of them in high black hats, one in an old opera hat, another in a pith helmet. Their clothes too were of every cut and material, and of the seediest description—but they will serve for years yet, as no doubt they are only worn on grand occasions. Poor fellows, thus attired they looked so mean and awkward, whereas on shore, barefooted, and in their native dress, they are such manly active fellows! All behaved admirably at service, for of course they could not understand a word, although two or three had brought their hymn books in the Kanaka language, and proposed to compare them with ours; but as neither they nor we could translate them, the comparison did not go very far. I do not know of what denomination they are, but their services are long ones, lasting over two hours, but are well attended. No buying or selling of any kind, not even of food, is allowed on a Sunday. No canoes are to be seen out fishing, and none come alongside to barter their little trifles as on week days. There is a fine of ten dollars on a chief seen drunk, of five dollars on an ordinary native. The importation or manufacture of intoxicating liquors is strictly prohibited, and a fine of five dollars a bottle imposed on any one breaking this law. The result of all this is, that you see a healthy, happy, and contented people, loving and respecting their government.

We had now visited nearly all the chief islands of the Marquesas and Society Islands. In the first of these groups, and in Tahiti and Morea, with the exception of Borobora, annexation, and French protection is complete—the consequence being that disease, drunkenness, and immorality are rampant. The French officials I fear, often do not exert much influence to check this condition of things, and the poor natives are dying out fast. The land, much of which was once under cultivation, is now abandoned for want of labour, the importation of coolies being prohibited, or subject to such endless restrictions that no one will run the risk and annoyance; in a word, the islands under French rule are falling from bad to worse. The natives have lost all respect for their own chiefs, who are powerless either to protect or punish them, and cannot but hate their French protector—invader? The few French settlers one sees are old time-expired soldiers or sailors, who having entered into relations with native or half-caste women during their period of service, prefer to take their discharge and remain in the country rather than return to France. The priests tell the same story, and even the government officials say that the system imposed by the home government is the ruin of the European settler. The gain to the actual revenue of France must be next to nothing, but they hold (with the exception of Ota Vanna harbour in Borabora) all the good harbours in the Marquesas, Paumutu, and Society groups. Should the Panama canal ever be completed, the high road between Europe and our colonies will hardly have an available harbour that could not be practically closed against English vessels; but these are matters that with broken treaties, and other slaps in the face given to poor Britannia, seem nowadays to awaken neither surprise nor indignation in the English nation.

After our service was over, the prime minister, with Mr. Blackett acting as his interpreter, came up and said he wanted a few words with me, so we adjourned to the poop, where he said that he had been told by authorised French agents, that the queen and government of Borabora must accept French protection whether they liked it or not, and that this was for their own good, as if they did not do so, they would be annexed by the Germans, a cruel people who would speedily exterminate them all; that if they did not willingly accept before October, the French admiral in a big iron ship with guns that could kill them even if they hid on the opposite side of the island, would come and compel them by force. This was what French officers had told them, and what Moe had been expressly sent over from Tahiti to tell her daughter, and what he (the minister) knew the people of Raiatea had been told, which had so frightened them that they accepted the protectorate they hated. He said also that they had 600 good men armed with muskets, and he knew that the French could not bring more against them. (Poor fellows!) English protection they were willing to accept, but that of France would destroy their religion and their laws; finally he summed up by putting these questions: "Would the Germans be likely to come and take them? Would the French admiral come, and if he did, had he any right to force the French protection on them? Whether Frenchmen were not all lies, and whether they would not be right in offering resistance?" To these questions I answered, that I had no official capacity whatever, and that consequently my answers would only be the opinion of a private person, but that I believed what was told them about the Germans was a simple fabrication. That the French had not only no right to force their protection upon them, but that there was a treaty with England by which it was undertaken that France would not interfere with them in any way; that this was a treaty *I hoped* England would make the French respect, and that if the admiral did come in October, which I did not believe he would, I should advise them not to resist by arms—for the French would be too strong for them; to accept no bribe, and to sign no paper of any kind, but to enter a formal protest against the French occupation on board the first English man-of-war that came into their harbour. The minister thanked me for what I had said, and told me he should advise the queen and chiefs to act on my advice, as he felt it was best for them, and the consultation having come to an end, the royal party went ashore again. Moe had not been allowed to come with them as she was favouring the French. On the following day I took the two boys ashore early in the morning to say good-bye to the queen, but the little lady was not up, so we went on to Moe's house. I took the opportunity of telling her how wrong I thought it was to advise her daughter to allow the French to establish a protectorate, when she knew how baneful that influence had been in the other islands, and how utterly it must upset and destroy the primitive, happy life of Borabora. She said she could not help it, as the authorities at Tahiti forced her to do so, and that she wished the islands belonged to Queen Victoria, who was a good Protestant queen. I had nothing more that I could urge, so we shook hands and said good-bye. On going back to the mole we found Mr. Blackett with a young turtle under his arm as a parting present. He said it would be useless to attempt to land at Tubai as the wind wouldn't suit, and that it might be some days before it would be feasible, so we had to give up the idea and let the tropic birds keep their tail-feathers untouched. Having said good-bye to this kind friend, it was not long before the boats were hoisted up, and we were steaming away from kindly little Borabora. At present it is a happy little kingdom, with a people leading a sober, righteous, and godly life; let us hope it may be spared from French protection, civilisation, and misrule.[1]

[1] Curiously enough, after this was written, but before it was in print, an article in one of the evening papers of the 22nd January, 1883, called attention to the points I have alluded to regarding the French occupation of Raiatea and threatened protection of Borabora, and I am glad to see that the writer echoes my own feelings almost word for word.

## CHAPTER XIV.

### *TONGA TABU.*

Aug. At noon we were thirty-six miles on our way, sliding comfortably along under canvas, on our way to Rora Tonga; during the next two or three days, however, we got such boisterous nasty weather, with rising sea, pelting rain, and fog, that when about twenty miles off Rora Tonga, we changed our course to go round to the leeward of the island, and shaped our course for Tonga Tabu, having had the satisfaction, during a lift in the clouds, to catch sight of Rora Tonga for a few minutes and ascertain that the estimate of our position was a correct one. The weather continued
Aug. very threatening and uncertain until the 21st, when we got
Aug. a pleasant change which lasted us to the 24th, when we sighted the island of Eoa, and shortly after the island of Enaigen, and steering between the two, stopped off Tonga Tabu about eleven, and signalled for a pilot, the chart saying one was to be had. We could see several natives on the beach, but no canoes, so we began to feel our way cautiously up the intricate channel which leads to the harbour of Nukalofa. As we were steaming in, a whale-boat which we had noticed beating out of the channel, crossed our bow, with the skipper of an American whaler on board;

he kindly offered to pilot us, and coming alongside got into the foretop, so as to be able to discriminate nicely between the different shades of green in the reef water, and keep clear of that which was of too light a colour, and handled the *Wanderer* as if he belonged to her. He was a romantic looking figure, giving the idea that he must have missed his vocation in not living with a rifle in some cavern in the Rocky Mountains, furnishing material for a book by Captain Mayne Reid. After about an hour's threading of the maze of reefs we anchored within view of Nukalofa, whose strongly constructed stone pier and farm-houses, showed a disappointing amount of civilisation. Unlike most of the South Sea Islands, with their fancifully shaped mountains, Tonga Tabu is as featureless as the West African coast at the Gaboon. One or two boats with natives soon came alongside. They were good looking and had very pleasant expressions, but their attitudes were so obviously studied, that they gave colour to the report that they are a remarkably vain people, and to the yarn of their having attributed the Duke of Edinburgh's failure to visit them, to his royal mother's fear of his falling in love with a Tongese lady, and subsequent illness when she refused to look at him. Their hair is plastered over occasionally with lime, which gives them the appearance of a London footman, when it is washed out again the hair is left of a yellow colour and stands out in a splendid friz.

After we had anchored, the professional pilot, a quaint old fellow, came on board; he had an idea that his dues should be paid whether he worked the ship into harbour himself, or merely assisted with his countenance from the beach. His English had evidently been picked up like that of many parrots from listening to sailors' talk, and the adjectives he used in telling Mrs. Lambert that he would bring her "plenty yam, fish and cocoa-nut" were possibly forcible, but certainly coarse. As soon as possible we all went ashore, and found the natives playing cricket on a very tolerable quality of turf. We called on Mr. Symonds, the commissioner, whose house, which is charming, is of native pattern, the outside walls having a sort of woven pattern of bamboo strips, and a thatched roof; inside, the house consists of one large oval room, with high open roof, most beautifully lined with matting and fibre; portions of the room were screened off at each end for dining and sleeping rooms. There were tables with curiosities and many pretty things about, vases of flowers, and books, and a garden with beds of rare shrubs; altogether a bright, refined home. After this visit we went on to pay our respects to the king and queen. The former is a very old man over eighty-six years of age, and at least six feet six inches high. He wore a black cloth round his loins and a blue flannel shirt, he shook hands with us and smiled, but as he understood no English, the interview was not a long one. The queen, whom we did not see, is said to be a splendid display of fat, weighing four hundred pounds, she cannot walk upright, but crawls about on her hands and knees. In the evening Mr. Symonds came off and dined, and told us of duck to be shot, and of an excursion into the heart of the island, for which he would provide horses. The following day some of our party tried to utilise the commissioner's information by going to a lagoon about two miles off after wild duck. Like other of our shooting expeditions this brought a great deal of vexation of body from the heat, thorns, and flies, but no sport. The fact is that it is very little use for passing travellers to try and get any shooting except in a few singularly favoured places. To make a successful bag you must have a knowledge of the habits of the birds and beasts, as well as of the peculiarities of the particular place and its climate. Tonga Tabu may be defined as an area of some three hundred square miles of forest and thick vegetation, intersected here and there with paths of mud and turf, and surrounded by reefs swarming with octopus, whales and sharks. Walking along through the forest paths, our ears were caught by a strange sound of muffled hammering. It suggested the possibility of some religious mysteries, but turned out to be nothing more romantic than the manufacture of *tappa* (the cloth the Tongans use for clothing), made by beating out the pith of a tree into the consistency of a sort of parchment.

This manufacture has been the occasion of a good deal of bitterness, for it did not at all suit the views of certain traders to Tonga Tabu, who suggested that one of the first signs of civilisation was an eager desire for Manchester cotton goods. Such strong pressure was put upon the old king, that a series of laws was promulgated, desiring the native population to attend church clothed in English stuffs, instead of the tasteful striped and chequered tappa dresses of their native land. The very making of tappa was prohibited, and the Manchester agency successfully established. The love of tappa, I am thankful to say, was however too firmly ingrained, and after two or three years of national martyrdom, the law has been repealed, and we found the island resonant with the renewal of the manufacture. Latterly the English Government seem to have realised the cruel wrong often done by British subjects on the South Sea Islanders in trading on their weaknesses, and are trying to remedy the mischief by appointing as consuls and commissioners men with exceptional powers and ability to undertake the responsibility intrusted to them. They can prevent the introduction of liquor; and if reasonable suspicion exists for supposing an Englishman to be acting dishonestly by the natives, he can be expelled from any given island. A most wholesome restraint has thus been established, which is appreciated by the natives, but which curtailing, as it often does, the gains of the white settlers, causes at times violent complaints to appear in newspapers whose circulation depends on the patronage of these men. Passing the groups of natives on shore, one could not fail to be struck with the pleasant relations existing between them and Mr. Symonds, created by his study of their manners and habits of thought, and shaping his conduct in accordance with these. "I always remember," he said, "amongst other things, that a chief here is as much deserving of consideration in his way, as a person of position in England; they are the first to recognise this, and appreciate one's attitude towards them. What business have we to suppose that our customs are essential to the well-being of those whose climate, country, associations, and thoughts, are cast in so utterly different a mould?" The more we saw of Mr. Symonds, the more we felt that we had at Tonga Tabu a commissioner, not only of exceptional ability and intelligence, but also a man who with a high sense of honour, has the rare gifts of sympathy with, and comprehension of, native character.

Next morning, all our party (excepting Mrs. Lambert, Miss Power and Willy) landed for an expedition on horseback with Mr. Symonds to Moa, the old capital of the island, where we were to find relics of Captain Cook, cyclopean monuments to past kings, caves with extraordinary stalactites, until lately unknown to any but a few natives; and also to have a taste of country life, feeding on Tongan dishes, and finishing up with a kava party. Mrs. Lambert watched our start from the ship, and it seemed to her a prolonged confusion of packing and saddling unbroken horses, whose attempts at bolting loose, sometimes necessitated a wild chase along the strip of turf stretching between the line of houses and the beach. At last the cavalcade disappeared behind the knoll on which stands the church erected to the memory of Captain Coker, who was shot down while assisting the missionary party in their endeavour to gain ascendency by force. About five miles of fairly good riding country brought us to the scene of Coker's fatal exploits. A certain mystery attaches to this unfortunate affair. The resistance to the universal introduction of Christianity by the king, whom the missionaries had gained over to their side, was organised, there is reason to believe, by certain beech-combing whites, who saw in the new *régime* an end to their nefarious dealings with the natives; and some evidence, although of a vague and imperfect character exists, tending to show that Captain Coker fell by the hand of his own countrymen, against whom, quite as much as against the idolatry of the people, his indignation had been roused. We found the village was surrounded by a deep and wide ditch, full of water. At the back was a strong stockade; behind this the natives had defended themselves with great gallantry against the invaders. The captain is said to have led his storming party with his sword in his right, and Bible in his left hand,

washing yourself. When this has been thoroughly done, carry off the bowl to the scene of festivity, and serve in polished cocoa-nut shells. After the careful preparations we had witnessed it was ridiculous to think of hurrying on the dinner. Luckily our minds were diverted by a chance remark of Mr. Symonds which set us hunting for makafechis—one of the strangest curiosities of the South Seas. This is a rat-shaped bait round which, when dangled in the water over the edge of the reef, the octopus wraps himself so tenaciously as to enable the fisherman to haul him out. There is a great deal of work in one of these baits. For the centre, a piece of quartz-stone, sometimes of an agate species, is rubbed into a perfect cone. This is backed by bits of mottled shell, kept in place by cocoa-nut fibre which passes underneath and extends past the point of the cone, into the semblance of a tail. There are one or two characteristic native traditions to account for the peculiar hostility of the octopus tribe against the rat tribe. They were formerly warm friends, but a rat, on a volcanic island that was suddenly found to be sinking below the surface of the water, having called in an octopus to carry him on his head to a more secure dwelling-place, with promises of cocoa-nuts in return for safe carriage, not only forgot to pay his passage, but having felt ill on the voyage behaved in anything but a nice manner; these facts so rankled in the hearts of the octopuses that they are quite unable to resist making an onslaught on a bait which combines the elements of both rat and nut. Unluckily for us the natives set great store by these makafechis, and by the time dinner was announced we had only secured some fragments of them.

On the floor of the hut, where the feast was to take place, banana leaves had been spread to serve as table-cloths, and upon this the viands which had been cut up into junks were placed—a limb of chicken, lump of pork, and great piece of snow-white mealy yam, in front of each person; in addition to this the chief's perquisites were set before myself, consisting of the pig's liver and the parson's noses of the fowls. We started very decorously with knives and forks, but found that our fingers were far more expeditious, which we cleaned afterwards by crunching out the juice from pieces of banana stems placed before us—a most excellent substitute for water. The banquet over, the cook was complimented upon the raciness of the dishes, and pipes being lighted, the humbler natives fell to on what was left uneaten. Soon after this Mr. Symonds and I took Nellie and Beatrice, who were getting very tired, to the house of a Danish trader, Mr. Neilson, whose kind wife provided them with a cup of tea and some nice cakes, and then showed them into a neat little room where she had provided clean tidy beds for them, and we then returned to the kava drinking. The ceremony begins by the head of the floor being taken by one of the chiefs, "the speaking-man," as they call him, or toast-master, who calls out the name of the person to be served, and to whom a cocoa-nut shell full of kava is handed; this has to be emptied at a draught and the bowl to be turned upside down. I was described in this fashion, as "Chief with the great ship," and one of our party who had not made any score at cricket as the "Duck Egg." Singing also went on that was rather pretty. Meanwhile the curiosity of the village was roused, and dark faces on the top of light dresses showed at the doorways or appeared like shadows in the dusk outside. At first the scene from its novelty was amusing. Kava, although to us tasting little better than soapsuds, has a certain exhilarating power, and the natives became cheerful in gesture and utterance. It has also a curious effect, for it stimulates without confusing the brain, and when taken to excess paralyses the lower limbs, so that the drinker finds himself unable to move, but with his mind singularly clear. Whether affected to this extent or not, it seemed as if our hosts had no intention of moving and were evidently in for a night of it. I went off to lie down, and found that kind Mrs. Neilson had got a sofa ready for me, and when I expostulated with her for taking so much trouble, she told me I was too old to sleep upon the floor; so after all I thought, old age has some advantages. The rest of the party lay down in the festal hut and tried to get to sleep, but it was not an easy matter. Saddles are very hard pillows, and packing on the floor with a lot of natives, skinful of kava, is not conducive to sleep in the tropics.

26th Aug I was awake and up by daylight, but, oh! the aches and pains, what with yesterday's ride on a rough hack, and the sofa, I could hardly move at first. Good Mrs. Neilson was up before I was, for on opening the door she appeared with three cups of tea with fresh milk and cake; nothing could have been more welcome, and having passed two cups into the girls room, whilst they were dressing I walked round to the chief's house, to see how the rest of our party had passed the night. On my return I found the two girls dressed and sitting before a capital breakfast of chicken and poached eggs, after which we took leave of our good hosts, having their promise that they would come on board next day to pay us a visit and see the ship. Whilst waiting for the saddling of the horses, we walked to see some so-called sacrificial altars, which we were assured were nothing of the kind, but in reality the burial-places of the queens of the spiritual kings of Tonga who resided at Moa. They are enveloped in a dense vegetation, the head-stones weigh from one to two tons, and are of a material not found in Tonga Tabu, but brought, tradition says, in large canoes, from a neighbouring island, where large quantities of this kind of stone still are to be found. From thence to the graveyard of the spiritual kings themselves, the graves headed by similar stones, but very much larger than the queens have.

At about 10 A.M. we started again under the kind guidance of Mr. Symonds, and rode about two or three miles to a part of the shore where the cliffs are highest, and where the infiltration of the water has eaten out the limestone into some remarkable stalactite caverns. The entrances, hidden by a luxuriant vegetable overgrowth, have been known to past generations of Tongese, but until Mr. Symonds won their confidence, they kept this knowledge to themselves. Dismounting, we found that the entrance the guides cleared for us was steep and low, and the attitudes necessary to pass into the lofty chambers within, decidedly ungainly and uncomfortable. It was just as well that the light given by the dried and platted palm leaf torches which the natives carried was dim and fitful, for if the depth of some of the fissures we skirted, and the precarious nature of the footing, had been brought into clear relief, no doubt that some of the party would have had their exploring enthusiasm considerably damped. Some of the chambers were magnificent in their natural architecture. Here and there the stalactites were so massive, regular and shapely, that little imagination was needed to fancy we were admiring a column of some grand old cathedral transept. The heat was excessive, and we longed to reduce our clothing to the standard of our Tongan

MOA CAVERNS, TONGA TABU.

guides; altogether, we were badly equipped for the enterprise. Riding-habits are not made with a view to clambering over dripping limestone rocks, and along rocky ledges, neither are stiff blucher boots. The struggle through the labyrinth of porch-like passages, sometimes so narrow that we had to push through sideways, to the sound of echoing voices, and the splashing of water falling from dimly seen heights, into unseen depths, was a strange experience, but what followed when we came into the last great chamber, was the most weirdly picturesque scene of our whole voyage. On one side the floor broke away into a deep pool many feet below us, into which poured all the waters of the various caverns. Fresh torches reserved for this moment made glare enough to bring out the roof and penetrate to the surface of the pool. From overhanging points of rock the natives plunged fearlessly, delighted, as we should have been, to refresh themselves after the hot tiring scramble. The dark bodies of the men as they bounded from the rocks into the eddying pool beneath, the hollow echoes of laughter and shout, the splashing and gleaming waters, and the cavernous chamber with its spiky roof, lit up by the gleaming of the torches, created the impression of some hobgoblin chamber filled with demoniac riot. Coming out after, and getting down on the shore, the gentlemen of our party managed to get a bath in the coral basins. The girls became the objects of a curious attention. Seeing their hands were covered with mud, the dripping divers stripped off the scanty garments that remained to them, to use as wet towels, and as they were unable from want of Tongan tongue to nip this kindness in the bud, they had to make the best of the civility.

We got back to the ship about 6 P.M., having enjoyed this two days' trip immensely, and found Willy full of excitement over a whale that had been caught the day before and brought alongside the American whaler; and what with

the steward ashore for fresh provisions. It had been our intention to make but a very short stay here, but we were told by the pilot that there was not a pound of coal to be had in Levuka or at Viti Levu, to which place the depot had lately been transferred. A vessel from Australia was, however, daily expected, so there was nothing to be done but wait, for we did not feel inclined to go on to Honolulu, some 3,000 miles away to windward of us, with only forty tons of coal on board.

1st Sept. The men caught a shark ten feet long, and got out of his inside two or three fish-hooks and some fowls' feathers—no wonder he was ready to take the bait. After breakfast all went off to call on the governor, Mr. Des Vœux. His house stands a little way out of the town, one storied, with a verandah, and pretty garden. We stopped to lunch in a dining-room tastefully decorated with tappa, paddles, and feather ornaments, and were shown two new rooms, just built for the young princes (who were daily expected to arrive with the flying squadron), ornamented with white pleated tappa, looking so fresh and nice. After lunch we went off with the governor to the reef, and examined the coral and fishes through a kind of box-tube; the same sort of thing is used by pearl fishers, as it enables them to see the oysters at a much greater depth than with the naked eye. In the evening we went ashore again, to dine at Government House. The dinner table was a most refreshing sight, with its wealth of beautiful flowers, white lilies, and tropical leaves. With the exception of an English butler, the servants were all Fijians, dark-skined fellows with hair dyed yellow, and dressed in collarless tunics of white cotton, with loin-cloths bordered with scarlet—a cool and pretty costume, which is in fact the uniform of the governor's body-guard. In undress they leave off the tunic, only wearing the loin-cloth, which reaches to the knee. I watched them at drill several times with admiration, fine strapping fellows, not less than six feet high, ready for work either afloat or ashore, and perfect water-dogs in the sea. I wish our English regiments could show men of equal physique. This little corps is composed I believe of only about 150 or 200 men, and does great credit to the commandant, Captain Herbert, who is also A.D.C. to the governor; he has one subaltern officer, and these two, with the addition of an officer who commands the whole rural police, form the entire military staff of Europeans at Mr. Des Vœux's disposal. The petty officers and men are natives; the outlying districts are governed through the chiefs. Nothing can be more orderly and quiet than the conduct of these islanders. The system which was organised by Sir Arthur Gordon, and does him the greatest credit, is most ably administered by Mr. Des Vœux, who rules this group of islands with the help of half-a-dozen Englishmen, and is developing its resources so that it may become self-supporting. One could not help contrasting the government and system with that of some of the French-ruled islands we had lately seen.

All off for a walk on shore, and to look about the town, 2nd Sept. which seems a busy little place. We met many English people, and found plenty of shops and two hotels. Shortly after our return on board, the governor paid us a visit, bringing with him the ex-King Thakombau, the veritable king of the cannibal islands. He is a fine old man, very upright, and not wanting in a certain dignity of gait and manner, quite white-haired and nearly blind. They say he has eaten hundreds of human beings, and the only favour he now asks of the governor is to be allowed to eat one more man before he dies. He wore a white shirt, with a piece of tappa round his waist, and bare legs and feet. His two sons came with him, both fine tall men, speaking English well, one of them having been educated in Sydney. I asked the governor what we should do to amuse the old king; he said he could not think of anything, unless I fired off some rounds from one of my Nordenfelt guns, as such an implement being more destructive of life than anything he had ever seen would be certain to delight him; this was accordingly done, and his Majesty seemed greatly pleased. After a glass of champagne (the only thing the governor would allow me to offer, as the regulations are very stringent with regard to supplying liquor to the natives) our royal visitors took their leave—going, as they had come, in the governor's state barge, manned by sixteen oarsmen in their bright white and scarlet dresses, making a very pretty picture. In the evening Mr. and Mrs. Des Vœux, and the staff, consisting of Captain Herbert, Mr. Amherst, and Mr. Allardyce, dined with us.

3rd Sept. This morning about 6 A.M. we were roused by a noise of shouting far away in the distance inland; it grew louder and nearer until it reached the town, and the whole place seemed to ring with it. Then it was taken up by every coasting ship in the harbour, and every canoe that held a native gave out the cry of welcome. Sending a man up to the mast-head, he soon reported the squadron in sight, and at about 7 A.M. we could see them from the deck. The excitement now was really intense and most infectious. Numbers of natives had been assembling for some days from other islands to welcome the young princes, and the water inside the reef was

FLYING SQUADRON, LEVUKA

covered as if by magic with great canoes crowded with eager people. The large picturesque sails, made of a sort of matting, were bordered down the leach, with banners and bannerols of every conceivable shape and colour, whale-boats, and cutters under sail also loaded with natives in varied dresses, with faces painted, some red, others black, contrasting with their brown and burnished-looking bodies, clustered around to do honour to the grandsons of "Marama," as our Queen is called. It was a beautiful sight, on which the sun shone brilliantly, and brought out vividly the contrast between the pale green of the water within the reef, and the deep blue of that without, with its white foaming crests, and lit up the gallant squadron fast approaching under canvas, until within half a mile of the entrance, when all sails were furled, and the ships threaded their way under slow steam through the narrow channel. The *Inconstant* (flagship) led the way, followed by the *Tourmaline*, and then the *Bacchante, Cleopatra*, and *Carysfort*. Every ship in the harbour was dressed, including the hulks, drums and fifes resounded, mingled with the sound of the wooden drums and sounding of conchs from the canoes. After lunch was over, and all our saluting done, I went off to call on the admiral and captains of the other ships.

Mrs. Lambert and most of our party went off to the *Inconstant* for service, finding friends there in Mr. Bigge and Captain Fitzgerald. They came back pleased with their visit—the men's singing—and the sermon. The Admiral (Lord Clanwilliam) had also showed them over his quarters. The afternoon was a busy one receiving friends and visitors—hosts of natives and foreigners came to see the ship, the former always delighted to look at the likeness of "Marama" 4th Sept.

which hangs in the poop-cabin. Many of our men met old shipmates, whom they had not seen for years, so that our Sunday afternoon was a very different one to what we usually spent. In the midst of all this gaiety, Beatrice was unfortunately laid by with a touch of low fever, brought on we thought by over-fatigue at Tonga Tabu.

5th Sept. After lunch we went ashore to the governor's to see the state presentation of the old king and his chiefs to the royal princes, the admiral and captains of the squadron. We were singularly fortunate to have been here at this time, for it is not likely that such a sight will ever be seen again. The ceremony took place on the green slope in front of the governor's house. The princes—governor—admiral—and naval officers in uniform were seated on mats, surrounded by crowds of natives and English. In front of them was a group of chiefs with the kava bowl, and on one side stood a native band who chanted a wild song, accompanied with the beating of hollow pieces of bamboo on the ground, making a noise which was far from unmusical. At each particular process of the kava brew, there was a fresh outburst of music. After it had been well strained with bunches of hibiscus fibre, and was quite ready, a chief decorated with long trailing leaves secured round his wrist with a broad bandage of tappa, his face painted in red and black patches, took a little of the kava, and presented it to the governor, admiral, the old King Thakombau, and the young princes. This ceremony used to have a semi-religious character about it, but I am afraid that Prince Albert Victor made a decidedly wry face as he took his portion, and that his brother laughed. After this there came marching up the centre a procession of natives, each one of whom carried some offering—which was cast down on the ground before the princes—such a collection of yams, fowls, turtles, and fruits! Having got rid of their burdens, the band played and the tribe burst into a furious dance—the same that used to take place before a cannibal feast. The men were dressed in the most fantastic way, their heads adorned with scarlet hibiscus flowers, or with great chignons of brown and white tappa—their faces and

FIJIAN DANCE.

bodies painted, and hung with leaves and fibrous plants, and long strips of tappa which trailed behind them. They bent their bodies backwards, then forwards, until their heads almost touched the ground; sometimes they lay flat on the ground, and yelled and shrieked in a manner that must have been anything but comforting to the unfortunate *larder* in the old days. Perfect time was kept throughout, then having

finished, they filed off to be succeeded by another tribe, who went through the same performance, until the offerings in front of the princes had grown into perfect heaps. It was a most striking scene, and perhaps not the least remarkable point about it was, that these wild savage-looking fellows were in reality a body of respectable and well-behaved Wesleyans. We were told that a French ship had lately been in, trying to learn our secret in the government of these islands; as I said before, there is not a single European soldier except the two or three officers; neither is there any ship of war stationed here. We went back to the ship for dinner, and returned ashore afterwards, to hear the band of the *Inconstant* play, and see more native dances, with crowds of lookers-on, townspeople, British sailors, natives, officers and ladies. It was a most beautiful night, and the natives were highly delighted with a display of the electric light from the flagship. When first seen it was greeted with yells of approbation, repeated as the rockets shot whizzing up in the still clear night. The only thing that damped our pleasure was Beatrice's absence, but it was thought better to
6th Sept. keep her quiet on board. We refreshed ourselves after the excitement of yesterday with a pleasant sail in the galley, beating round the island inside the reef, and through intricate narrow passages of coral which are growing up everywhere inside the reef, and unless blown up, will seriously choke the harbour. In the afternoon there was another meki-meki, as these dances are called, and in the evening the captains of the squadron and flag lieutenant dined with us. At night there was a ball at Government House—the large
7th Sept. room well lighted with Chinese lanterns. The verandah was a cool, pleasant spot for chaperons to look on. We

were surprised at the number of ladies, but they came from various parts of the island, and some even from Melbourne. What enterprising people they must be in the colonies! Nellie had the honour of being presented to Prince Albert Victor, and waltzing with him. We left about 2 P.M. and went off to the ship without any wraps—it was so warm and still.

8th Sept. The excitement was well kept up during the remainder of our stay in Levuka. For to-day the admiral gave an afternoon dance on board the *Inconstant*—the ship beautifully decorated, the bandsmen in white, and the visitors all in their best clothes. Numbers of Fijian chiefs with their wives and daughters squatted on each side of the deck, and added much to the gaiety of the scene. The women dressed in wrappings of tappa, with a little tight bodice of scarlet, blue, or white. The men with flowers in their hair, and necklaces. They applauded some of the dances, calling out "Vanaka!" "good;" but I should think they must have seemed very tame performances after their meki-meki. In the evening some of our old naval friends dined with Mrs. Lambert, whilst I went to dinner at Government House, and met the admiral and senior officers of the squadron, after which more meki-meki.

9th Sept. To-day we had a cricket match, and after this a final meki-meki to lay farewell offerings before the princes. Tappa, war-clubs, mats, fruits, vegetables, poultry, turtles, etc., such a collection! Then came kava drinking, winding up with a most extraordinary dance supposed to be descriptive of a search for a goddess, followed by the women's devil-dance—a mixture of war-dance and pantomime carried out with extraordinary energy of limb and gesture. At night

[illegible]

REWA RIVER, FIJI.

with a present of clubs and mats from their father, with many kind and civil speeches.

On Sunday morning, at 6 A.M. we left Levuka, and had a nasty, rough passage to Suva, the capital of Viti Levu, and the future seat of government. We took several passengers with us, European and native, who had been to Levuka for the gaieties. Amongst them being Ambarosa, a nephew of King Thakombau and a chief of Viti Levu. We arrived in Suva harbour at 2 P.M. It is much more roomy than Levuka, and the shore is low and flat, running back for some distance, admitting of a town being built, which could not be done at Levuka to the same advantage. The entrance through the reef is also clean and well sheltered in any weather. I went on shore to pay a visit to Mr. Seed, the only political officer at present in Suva. He is superintending the construction of all the new government buildings, many of which were in a forward condition, and when finished, the seat of government will be shifted to this island. 11th Sept.

In the morning the steam-cutter was lowered and steam got up, and all necessaries for a two days' expedition got ready for ourselves as well as the crews of the cutter and galley, as we had in view an expedition up the river Rewa, to get an idea of the interior resources of Viti Levu, upon which the colony must chiefly depend, as it is by far the richest and largest of the Fiji group. At half-past ten we were joined by Ratu Aboroso, or Ambarosa as he is generally called, our passenger from Levuka, and we started in the galley, in tow of the steam-cutter, with a stiff breeze blowing right in our teeth. The ladies had talked previously of coming with us, but changed their minds when they saw the pitching that we were evidently coming in for, and they 12th Sept.

were wise, for we soon got into a nasty, chopping sea, having the tide with us, and a strong breeze against us, through which we had to peg for about two hours before we reached one of the entrances to the mouth of the Rewa. In the roadstead

RATU TIM.

lay a fine iron barque, the *Peebleshire*, discharging machinery for a very large sugar-mill in course of erection on the river bank. It was low water, but a red buoy marked the entrance to the river, and round this we entered the narrow channel between two rows of poles, with wide mud-banks on each side of us. We passed numbers of duck and blue and white cranes, but refrained from shooting any, as we were anxious not to delay, but to get on and as high up the river as possible before dark. Winding our way, under the skilful guidance of our friend Ambarosa, between sedge and mangroves, for something like six miles, we came out into the main stream, a noble river, just below the native town of Rewa, which stands on the bank on some table-land, about fifteen feet above the level of the river. The side from which we emerged was still a mangrove bank for another mile or more, when it rose to the level of the opposite side, thickly covered with timber. A little above the town stands the Rewa hotel, a neat-looking house with a nice lawn in front of it, backed by cocoa-nut and banana trees. As we passed, a red ensign was run up to the flag-staff, as a compliment, no doubt, to the St. George's cross flying at the galley's mizzen. Soon after passing this spot a canoe came out of one of the many creeks, and, as it made for us, we stopped and bought a couple of dozen cocoa-nuts, for which we gave sixpence to the old fellow in the boat, much to his delight. A bend in the river now enabled us to set our canvas and help the cutter's engine, and we bowled merrily along, passing, as we ascended, many fine and thriving sugar plantations, and a sugar-mill, nearly completed, with a jetty at which were moored a large steam launch and some iron lighters. Then more sugar plantations and many islands. Finally the very large mills under construction for a company, called, I think, the Colonial Sugar Plant and Plantation Company, of Melbourne. They have, I believe, large establishments in Queensland also, and are said to have acquired large tracts of land on the Rewa at different points. Their machinery is capable of crushing twenty-five tons a day, and has cost £150,000; they employ a hundred white mechanics and several hundred natives and coolies, and they have a little fleet of four or five steam-launches and a number of iron lighters, carrying over 100 tons each. The manager very kindly showed us all over the premises.

After leaving here the country became more undulating, and the stream began to narrow, until about 4 P.M. we reached Mrs. Thomas's hotel, which we found must be our halting-place for the night. It stands on the point where the Wrinamu river falls into the Rewa, and is a very clean, nice little house, bounded by sugar plantations on three sides and by the Wrinamu on the fourth. We landed to see what prospect there was of Mrs. Thomas being able to provide food and lodging for such an unexpected influx of visitors. We found her most kind and obliging; she showed us into nice clean bedrooms, sufficient for all our party, and some to spare for some of the men; the rest she promised to make as comfortable as possible. Dinner also was promised us at six o'clock; so having settled all these matters comfortably, we took to the boats again and had another hour's steam up the river, passing constant plantations and picturesque native villages, with planters' residences. At 5.15 P.M. we were abreast of a most beautiful island in mid-stream, with hills on each side, and in the distance rose the blue-peaked outline of the highest range in Viti Levu. It was a most balmy evening, and we stopped reluctantly to retrace our way to the hotel, which we reached at dusk. Whilst the men were anchoring and securing the boats, we began to get ready for dinner, soon announced by good

Mrs. Thomas, and to which we sat down in the humour to enjoy it, as, indeed, we did, for no one could have wished for a more excellent and well-cooked meal, with capital lager beer, and such a pudding to finish off with—made with fresh eggs and rich cream—a pudding that those who wander on the sea know how to appreciate. Afterwards we went up stairs, and sat in a verandah that ran outside our bedrooms, to discuss the day's work *and* our baccy. No need to enlarge, O brother smokers, on the perfection of that smoke! A long day in the open air, a good meal, a pleasant sense of fatigue, a cool verandah, a still calm night, the bubbling sound of the river, the prospects of a good bed, what surroundings for a smoke! Beneath us the crews of the two boats discussed their dinner (which my coxswain told me was similar to our own) with evident satisfaction, as afterwards evidenced by the cheery songs and choruses which resounded through the night, until gradually, one after another, we sheered off to bed, and left the little hotel in peace and quiet.

We were up early after a good night's rest. Some of the party 13th Sept. went off to visit the native village before breakfast, which Mrs. Thomas got ready punctually for us at seven o'clock. We must have ascended the river in all about twenty-one miles, and Ambarosa told us we could have gone about five miles further. I should certainly have done this but for my promise to Mrs. Lambert to be only one night away; from which this moral should be drawn—never promise that you will return on a given date when exploring a new country, for by doing so we lost a great treat in not seeing a few more miles of the river of which we had evidently arrived at the most interesting part. Our breakfast was as good as our dinner of the previous night. And now for the bill!

Lodging, dinner, and breakfast for sixteen persons, amounted to £6 2*s.* 0*d.*! Think of that, ye harpies of European hotels! At eight o'clock we were in the boats again, and bidding good-bye to kindly Mrs. Thomas, who, we found, came from Melbourne, spun rapidly down stream, anchoring at noon to lunch, and after an exciting race between the steam-cutter under steam and sail, and the galley, under sail, came alongside the *Wanderer* at 3 P.M., after a most enjoyable excursion. All were well on board, but the coaling was not over, as we were stuffing the bunkers as full as they would hold, besides running a day's coal into the stoke-hole and taking ten tons in sacks on the decks, in order to face the 3,800 miles we had to go dead to windward to get to the Sandwich Islands.

## CHAPTER XVI.

*AN OLD CANNIBAL—FROM FIJI TO HONOLULU.*

COALING was over before noon, and after dinner all hands were busy in cleaning the ship after this necessary but nasty business. During our absence up the Rewa, Mr. and Mrs. Seed had dined on board, and on our return, Mrs. Lambert and I went ashore to bid them good-bye. To-day our good friend Ambarosa brought his wife and sister to see the yacht, and everybody was charmed with them, they were such perfect ladies, with the most gentle manners, soft voices and pleasant faces; they looked about, and made little sounds of admiration like the cooings of a dove, with occasional outbursts of "Vanaka, vanaka!" The pantry pleased them particularly, as did the arrangements for letting the sea-water into the baths in the cabins. One of the ladies wore a strip of blue-and-white print, bound round her with tappa, the other a red one. 14th Sept.

Ambarosa was a great swell, in clean white shirt with silver studs, black neck-tie, flannel girdle secured with a black scarf, and white turban, and glossy legs and feet. The ladies, before going, wrote their names in our visitors' book very well, although they were unable to speak English; one was Salate, the other Kelera (query Clara). They went off in their large canoe, and shortly after sent Mrs. Lambert a present of a basket and two skirts made of hair-like fibre. Ambarosa having escorted them ashore, came back in his very handsome canoe and took some of our party, with the two boys, out for a cruise. They got back to lunch, having been to the reef and taken up some fish-pots, the contents of which furnished us all with a capital dish of fish. In the afternoon we went out for a sail in the galley, taking Mrs. Lambert with us, and as we were cruising about, met Ambarosa in his canoe, and got him to come into our boat and pilot us up a small river there is at the mouth of the harbour. As we were going along Mrs. Lambert asked him if he had ever eaten human flesh. To our great relief he said No, but that he had seen it cooked and eaten many times, that the fat is yellow and smells like roast duck, and that the prime parts are the arms and thighs. It is but five years ago since this country was cannibal. The river was a very pretty one, full of fish, and with trees growing down to the water's edge, which were swarming with parrots, doves, and other birds. After about three miles we arrived at a native village, where some of Ambarosa's people lived. The boys landed with him, and going into the village, came back to the boat, bringing with them an old chief, who, Ambarosa told us, had clubbed and eaten many men in his day. On our asking him if this were so, he answered, smacking his lips and rubbing his old stomach, "Yes, many, in old time; very good!" This dreadful old man gave us a curious little wooden dish, something like a very flat, shallow sauce-boat, in which many good bits of humanity had been placed nicely roasted. We had a dead beat back, and did not get to the ship until dusk. Mr. and Mrs. Emprose, whom we had brought down from Levuka, dined with us in the evening; and as he was the representative of the firm from whom we got our coal, we were able to settle up all our accounts after dinner, and so clear the way for sailing on the morrow.

The last supplies of fresh bread and meat were got in this morning, after which we left Suva, and on getting outside the reef found we had to contend with a good stiff south-east trade, which would most likely last us for a couple of days, during which we should be clearing the Fiji group. We left this crown colony very reluctantly, having met with such universal kindness, and seen just enough to make us understand how much of interest and beauty we must be leaving unseen. The governor was going to start in a few days on an excursion right across Viti Levu, and had very kindly pressed me to be one of his party. It was to be a purely official tour, across the high part of the island, with the object of visiting the tribes that have only lately been subdued and reduced to submission. Until about two years ago they were a troublesome lot, making raids into the plains, fighting with the lowland tribes, and carrying off, when they could get them, victims for their cannibal orgies. It would have been a most interesting trip, and one would have seen a country and people almost unknown to white men. Travelling in these islands makes one understand why it is that such men as Sir Arthur Gordon and Mr. Des Vœux take such an intense interest in the work of government. In this fine but naturally unruly people they have a perfectly wild material to work on. Hitherto they have lived but for war and mutual destruction, their feuds excited and their warlike feeling stimulated very often by 15th Sept.

the selfish and vicious counsel of white men, often outlaws themselves, and whose lives of riot and debauch furnished an example only too readily followed by the poor natives, who, inflamed with liquor supplied them by such fellows, were ready for any deed of atrocity.

In the islands governed by the French, where such practices have gone on unchecked, the native population is being rapidly reduced in numbers—so rapidly, that one can easily foresee that in a few years hence their beautiful island homes will know them no more. Without these evil influences, and under a kind and just rule, these islanders become perfectly docile and friendly, respecting and loving the white man; and it certainly must be a work of enormous interest to gradually Christianise, civilise, and thus preserve, instead of destroy, this splendid race of men. The great difficulty to get over is their idleness, or rather their aversion to work; and one of the chief objects in the crown legislation is to try and induce the people to get over this aversion, in their own interests. They are taxed to cover public works, government and police, but instead of money, if they prefer, they may substitute sugar-cane grown by themselves; and this is gradually being done, a contract having been made between the government and a gentleman (who paid us a visit, and who has nearly completed some sugar works, that we saw on the Rewa) who takes all the government cane obtained in this way, as payment of taxes, at a fixed price, based on the current value of sugar during each successive season. A law has also been passed forbidding the sale, by the chiefs, of land they rule over, but power is given them to grant leases, subject to the consent of the head men, and approval of government. The reason of this law, which at first may sound somewhat arbitrary, is a sound one, for it is in reality for the protection of the natives and their property. Grants of large tracts of land are claimed by white settlers—a large proportion of which were no doubt obtained from the chiefs, ignorant of what they were giving, or when under the influence of drink—the consideration, a keg of rum, to which in some cases was added half a dozen old muskets and a little gunpowder. The chief, however, in reality had no right to make these grants, as the land did not belong to him individually, but to him and his tribe To clear up these claims a Royal Commission was appointed, with Sir Arthur Gordon as president; and sincerely do I hope that many of these unrighteous claims will be disallowed. In fact, the whole tendency of legislation is to protect the native, the original owner of the soil, and to try and civilise him. It is needless to say that with a certain class this policy is perfectly execrated, who consider that it is a Briton's right to ride roughshod over what they are pleased to call the nigger or savage—save the mark!—the savage being in nine cases out of ten a superior being to those who thus designate him. I cannot imagine a more hard-working lot of officers than the Governor of Fiji and his staff; zealous in the defence of the natives, yet keen in promoting and developing the resources of the colony. All honour to them! All this time we are encountering a strong breeze and high head-sea, and the ladies, and even some of the gentlemen, have to own it too much of a good thing, and disappear below.

We have had two fifteenths this month, and eight 16th Sept.
days this week, to make up for what we lost in crossing the meridian 180° longitude going west; while to-day we

left but yesterday to return to Chili, so that we just missed sending our letters to Janet. He gave us news of the sad assassination of President Garfield. One of our party asked him if Mr. Gladstone was still alive, to which he replied, taking off his hat and bowing, "He is alive and cock of the walk still, sir," the pilot being an American and an admirer of Mr. Gladstone. No sooner had we anchored than a boat came alongside from Mr. Davies, the vice-consul, bringing us a batch of welcome letters, with good news of all at home, as well as newspapers, which were eagerly devoured. In the afternoon we went ashore, and were struck with the busy, thriving look of the town. There were a good many vessels busy loading or discharging cargo along the quay side, and the town was much larger and more business-like than any we have seen for a long time. Open cabs, called expresses, with good horses, either of English breed or imported from California, were to be had. The streets reminded us somewhat of a London suburb. The shops, mostly kept by Americans, were draped with festoons of black and white, in mourning for the President. My wife and I went first of all to call on Mr. Davies, and thence to our commissioner, Major Wodehouse, to whose kindness we owed the early receipt of our letters. From thence to see the Bishop of Honolulu, to tell him of the special object of our visit, and ask his advice upon the matter.[1]

We drove through pretty green lanes to the bishop's house, which stands just out of the town, and found the "palace"

[1] Mr. and Mrs. Lambert's relations and friends are of course aware of the object here alluded to; for others who are not, it will be sufficient to say that their son Charles, twenty-four years of age, was drowned at Hawaii when bathing in November, 1874. No opportunity had offered of

to consist of a pretty cottage with creepers trailing over the verandah. He was everything that was kind, as well as his sister, a very charming person, who helps him in many ways, especially with a large boarding-school which he has adjoining his house, for native, Chinese, and half-caste boys. Unfortunately for the bishop, Miss Willis was shortly going to be married to a clergyman in Hawaii. After this visit was over we went on to pay our respects to Queen Emma.

8th Oct. It rained heavily up to eleven o'clock, and at one Major, Mrs., and Miss Wodehouse came off to lunch. At 3 P.M. the bishop and Miss Willis came on board, and in the evening we went ashore and sat under the trees in Emma Square, listening to an excellent band, composed entirely of native musicians, but conducted by a German bandmaster. The mosquitos on board at night were very bad, and even Mrs. Lambert was tempted to wish we were at sea again. On Sunday we had a very quiet day, a few friends coming off to service in the morning, and various particularly well-behaved and orderly visitors in the afternoon. 9th Oct.

10th Oct. On Monday we left the ship at 10 A.M., and had a busy day—calling first on the governor, an American gentleman, who has married one of the king's sisters. Then on the Minister for Foreign Affairs, both at their offices in Government House, a large and spacious building. We then went to see the studio of an American artist, who has made many interesting sketches and paintings of the lava overflow in

placing any memorial over his grave except the simple wooden cross which had been erected by their friend Admiral Cator, when Captain of the *Scout*. To replace this, a granite cross had now been brought in the *Wanderer* from England.

Hawaii, and which so nearly destroyed the town of Hilo. On all sides one could not help being struck by the patient industry of the Chinese; they were toiling away in all directions, reclaiming boggy ground, turning it into neat gardens, cultivating fruits and vegetables, keeping shops, acting as porters, cab-drivers—in fact, you might fancy yourself in China; the natives of the country seemed to be nowhere. We did meet a few women riding astride their horses, and one or two with the pretty flower necklaces, but they seemed more like scattered foreigners. The fish-market, described in books we had read, disappointed us, for the brilliant display of fishes of all shapes and colours was wanting; perhaps it was a bad day, or more likely our own experiences of the various reefs had shown us better things. One very sad and depressing sight we saw—the departure of a band of some twenty or thirty lepers to Molokai, the leper's island. There were men, women, and children, who, standing on the deck of a little schooner, waved their handkerchiefs to their relations or friends on shore, who could never hope to see them again, unless they followed them to Molokai, for they never recover or leave the island again when once they have landed. In the afternoon the younger members of our party got some horses and rode out to the Pali, getting a splendid view of the country from the top of the precipice, whilst we took a very pretty drive to Waikiki, passing rice-fields cultivated by Chinese labourers, and pretty country-houses.

11th Oct. Next day we landed soon after breakfast to meet Mr. Cleghorn (the king's brother-in-law)—who had kindly obtained permission for us to visit the royal mausoleum—a
12th Oct. rare favour, and one not granted since the visit of the

*Sunbeam* party. It stands in the centre of some garden ground, and looks like a little chapel. Inside it is very handsome, with a fine roof and beams of polished wood, and contains about thirty coffins, with the remains of all the Kamehameha line of kings, with the exception of the first and greatest of all, who conquered the whole island group, and formed a single monarchy for himself and his descendants. No one knows where he was buried, but it is supposed that he lies in the island of Hawaii, from whence he came. The coffins not occupied by the bodies of kings contain those of their queens, or of the princes who died before ascending the throne. We were struck by the enormous size of the coffins, which tell of the mighty stature of these island kings. The coffins are made of woods, both dark and light, indigenous to the country, and beautifully polished. Silver crowns and ornaments are fixed on each coffin, with the name of the occupant, and staffs, with the royal yellow feathers, emblems of royalty, arranged in large plumes, are placed around the coffins. Having listened to a brief history of the lives of each of the kings, we came away, much impressed with the reverence and care shown to these former rulers.

Later on in the day Mrs. Lambert and I went to call on the Princess Ruth, who most kindly sent her carriage (with coachman and footman in green livery) to meet us and bring us to her house, which is a little way out of the town. Princess Ruth is very decidedly inclined to *embonpoint*, but she has a very pleasant, kind expression of face, and we met her with peculiar interest, as she remembered our son Charles well. She has a nice house and garden, but a new palace almost finished is being built for her, just opposite Major Wodehouse's house. Here also we met Mr. Kaai, a native gentleman, who had been on the beach at the time of the accident, and had done all he could to be of help, getting the canoe which had brought our dear boy ashore, and helping to try and resuscitate him. It is needless to say with what painful interest we listened to all he had to tell us. On leaving, Queen Emma, who was also present, asked us to go for a drive with her, which we did, going first towards the Pearl River, and then up the Pali Road; and were much struck with the evident love and reverence displayed towards the Queen by all the natives, without exception. Indeed, there is no doubt that she is much respected by all classes. The Queen spoke much of the kindness she had received in England, saying she would like to go again. On our way through the town she stopped to get us some flower wreaths, two of which she put round Mrs. Lambert's neck, and one round my own; one was composed of small roses and ferns, another of a deep yellow flower and white jessamine. The flowers were beautiful. There was a hibiscus with a large double flower which is pure white in the morning, by about 11 A.M. it gets a faint tinge of pink, deepening throughout the day, until by evening it is quite red, and then it dies. What a theme for a poet! During our absence the young people had taken horse again, and accompanied by Major Wodehouse, and some of his children, had gone off for a long ride towards Diamond Head, to visit an old battle-field. When we got back to the ship we found to our great delight that the coaling was over, and all hands busy cleaning.

13th Oct. This was a busy day—receiving friends from shore,

amongst them Mr. Green, the Minister for Foreign Affairs, Mr. Dominis, and Mr. Cleghorn, the king's brothers-in-law, Mrs. Dominis (the princess unfortunately was suffering from an accident, having been thrown out of the carriage, and could not come), Major Wodehouse, and others, who stayed to lunch. Later in the day Queen Emma, with Princess Ruth, three ladies in waiting, and Mr. Kaai, came on board and spent the afternoon, having tea on the poop. Princess Ruth is the last of the old dynasty. She is a most good-natured, jolly creature, and seeing that Mrs. Lambert was looking over the side, as the royal party were putting off, she pointed to me, and then put her arm round me in a most protecting kindly manner. Major Wodehouse dined with us in the evening, after which the ladies went ashore to a bazaar which was being held in aid of the fund for building an English church for the Chinese.

On the 14th we went for a picnic to the Pearl River, 14th Oct. about seven miles from our anchorage. Our party was reinforced by several ladies and gentlemen from shore; we went in the steam-cutter, towing the galley. The harbour at the Pearl River would make a better one than that at Honolulu, but for the narrow entrance, and shallow water on the bar; once inside, it is a charming sheet of water. We were told that the United States Government had been treating for the purchase of this bay, with the intention of deepening the entrance, and making it their naval station in the Pacific, but that the negotiations had fallen through owing to the objections of France and England. We steamed round the jagged shores and islands, which help to form many a quiet little harbour, and remarked how everywhere there were signs of increasing prosperity, and a development of agricultural enterprise, in the shape of sugar and rice plantations. We landed and lunched on a pleasant island shaded with algarob and espino trees, and, after a couple of pleasant hours, returned to the ship by sunset.

15th Oct. To-day Mr. Cleghorn took us to see the hospital—a fine, airy building, well suited to the climate, and standing in very pretty grounds full of tropical trees, forming avenues of cocoa-nut, date and imperial palms, besides flowering shrubs and creepers of various kinds and colours. From here we went to the gaol, which reflects the greatest credit on the government; it is beautifully kept, and the exercise yard is planted with orange trees. Princess Ruth had most kindly invited us to come to an early dinner, which was to be cooked and served in the native fashion; and on arriving we found a large party, including Queen Emma. We were decorated in the usual graceful fashion with wreaths of flowers, and taken to the banqueting hall. Here we found a long low table, about a foot high, spread with a cloth composed of banana leaves, and decorated with bouquets, and covered with every variety of native dainty and luxury. I am sorry not to have a bill of fare to give, but I know there was cooked, dried, and raw fish, chickens, baked pig, game, raw shrimps, crabs, octopus, bowls of poi, boiled taro, sweet potatoes, water-melons, and many other good things. We sat on mats upon the ground, whilst children kept the flies away with feather brushes. The room was hung with garlands, and the table was a very pretty sight. I sat by Princess Ruth, and enjoyed myself exceedingly, finding that raw shrimps were most capital things. Mrs. Lambert sat near the Queen Dowager, who was most kind and attentive to her, being much amused at her feeble efforts to eat poi, and showing her how to eat it—for it takes the place of bread, and forms the daily food of the natives. At the gaol we had seen large bowls prepared for the prisoners. After the meal was over, we sat in the balcony, and listened first to the capital Honolulu band, and afterwards to some native music, performed by three men and four women dressed in yellow and white, who played on large gourds. Altogether it was a most pleasant, friendly entertainment, without any formality, and we got back to the ship delighted with our day.

Major and Mrs. Wodehouse and Mr. and Mrs. Feer with their two daughters came off to service in the morning, and we had some friends to dinner in the evening. 16th Oct.

On Monday we drove over to Mr. Cleghorn's house at Waikiki, and found a large party assembled to celebrate the sixth birthday of his little daughter, a very pretty, charming little creature, who is the only scion of the reigning family, and the hope of the Hawaiian people. Her mother, Princess Likiliki, we found most kind and attentive to us, as indeed was everybody at our host's. After a similar banquet to that at Princess Ruth's, we wandered about the beautiful grounds, and were shown a tree which our good friend Cator had planted, and were then entertained by a native song and dance called a hula-hula until the banquet had been cleared away, when dancing was carried on with great spirit for the rest of the afternoon. There were so many pretty girls that we could well understand why Honolulu is such a favourite place with naval officers. 17th Oct.

To-day we went to the cathedral to see the marriage of Miss Willis, the bishop's sister, to the Rev. R. Wainwright, 18th Oct.

19th Oct. and on the following day drove out to pleasant Waikiki, to leave cards on Mr. Cleghorn and the princess, and then to visit Queen Emma and Princess Ruth, whom we found with their attendants, sitting in the garden, and after a most pleasant interview came back to the ship for dinner, and to prepare for our start in the morning.

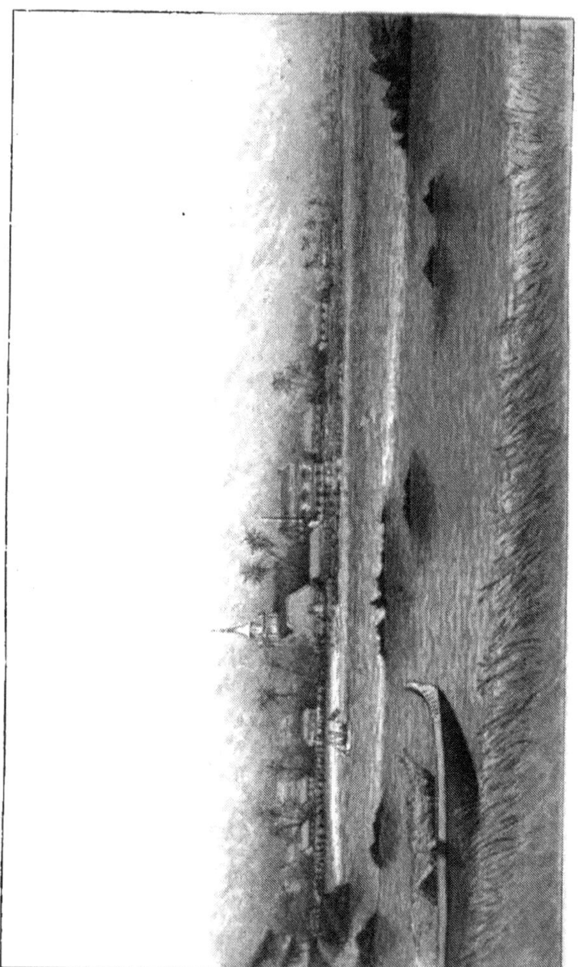

KAILUA HAWAII.

## CHAPTER XVII.

*KONA.—HILO.—THE VOLCANO.—KILAUA.*

AT 10 A.M. having taken Major and Miss Wodehouse on board we steamed out of Honolulu harbour and shaped our course for Cook's Bay, in the island of Hawaii, not, however, without having first received a pretty present from Queen Emma of flowers, honey and prawns, and of fresh eggs from Mrs. Wodehouse. We had a fine passage, and a nice clear night, and at 7.30 A.M. we passed close to Kilaua Bay, the scene of our sad loss. At ten we anchored in Cook's Bay, and as soon as the men had finished their dinners, the stones we had brought so far with us were placed in the boats and landed, ready to go up to Kona in the ox-waggons procured for us by Kavanah, the mason engaged at Honolulu. In the afternoon we landed, and spent some little time in a house belonging to Princess Likiliki, which she had kindly placed at our disposal; nearly opposite to us was the obelisk, erected to the memory of Captain Cook, almost on the spot where he stood, before receiving his death blow. The cliff rises from the shore perpendicularly, to a height of nearly 600 feet, its face honeycombed with caves, the burial-places of former chiefs. The people must have been lowered from above with the bodies, to place them in these very secure resting-places.

20th Oct.

21st Oct.

22nd Oct. At 8 A.M. we landed, with thirteen of the crew to help in loading the waggons, and to accompany them on foot, and help them up the steep road to Kona. The road for the first two miles was a steep climb over cinders and broken bits of lava, and through a most desolate country, with nothing to

THE GRAVE AT KONA

be seen but slag and scoria. After wearily ascending this road for some three miles, we came to some stunted vegetation, grass and scrub, which improved as we got further on, until we reached a pretty flowery lane, shaded by paradise and blue gum-trees, which brought us to the little church with its quiet graveyard. It is very beautifully kept, and without wishing to enlarge upon what can be of interest to a very few, I must record our gratitude to the incumbent Mr. Davies and his wife, for the care and attention shown to our dear son's grave. We were told by the neighbours, that for years fresh flowers had been constantly placed upon the grave and hung on the plain oak cross. Having seen the stones taken out of the cart, we retraced our way to the ship, passing a quiet day on the Sunday. 23rd Oct.

From Monday to Wednesday we were busy finishing the work which we had come for, and on the latter day had the satisfaction of seeing everything completed, and bade farewell with full and grateful hearts to Mr. and Mrs. Davies and the quiet resting-place in Kona churchyard. 24th—26th Oct.

From early morning we were steaming along a lovely but precipitous coast cut into with deep gorges and ravines, but studded with houses, cottages and sugar-mills. The sugar industry is the chief resource of the islanders, and promises well, one of the great advantages being that there exists a treaty with America, opening the American markets to this produce free of duty. On all the principal islands of Hawaii, land is being cleared, and there seems to be a promise of a prosperous future for the kingdom of Hawaii. At 2.30 P.M. we anchored in Hilo, and found at anchor the Russian Corvette, *Vestnik*, carrying a rear-admiral's flag. Lieutenant Polevoy, a very courteous officer, boarded us, offering his services; he told us that the admiral, with ten or twelve officers, had started the day before for the volcano, but were expected back on the following day; he invited us to visit his ship, which we did in the evening, being received with the band playing "God save the Queen," 27th Oct.

and entertained most hospitably. The band played on the deck, after which the sailors sang some Russian peasant songs. Most of the officers spoke more or less English, and showed us every attention and kindness.

28th Oct. To-day we went to visit the falls near the town, and see the diving. A native jumped from the cliff 100 feet high into the pool below, he took a run and came down like an arrow; others, both men and women, jumped from lesser heights, or sitting on the top of the falls allowed themselves to be carried down with the rush of water to the depths below. After watching them for some time we went further on for about two miles to the Rainbow Falls. We stood on high ground, beneath us was a huge gulf, on the opposite side two rushing streams of water, which joined just as they left the top, and fell 100 feet into a pool below, the spray rising in clouds, whilst behind the falls we could see the mouth of a cavern, overgrown with ferns. One side of the basin into which the water fell was a steep rock, but on the other side the ground was sloping, and covered with a mass of giant ferns and plants of various beauty. As we came back we called on Mr. and Mrs. Severance, who have a very pretty house and garden; they have utilised the water most cleverly, a constant stream flows through their bath room, and also works a sewing machine. The whole island of Hawaii seems as if it was the recent creation of volcanic agency—layer after layer of ashes, pumice, and lava, being spread one over the other; but this soon decomposes, and forms a very productive soil. The great drawback is no doubt the volcano, a most uncertain and unruly feature; only a fortnight before our visit he had shaken down a good deal of building with an earthquake; a few years before he devastated a whole district with lava streams, which, flowing into the sea, added a mile of promontory to the coast. Lately he again began to pour forth his molten entrails from a crater some fifty miles from Hilo. For eleven months the stream flowed on slowly but unceasingly. Those who watched it describe how, after spouting with a fiery rush down the steep descent, it crept along the hollows in the gentle slope of the mountain side, congealed on the surface with a hard black crust, but two or three inches down a glutinous mass of blood-red fire. Every now and then the stream would pause, and the end become completely caked over, and the people would think that its force was spent. Time after time, however, the liquid increased in volume inside, and bursting through the crust, stole on again for a further space, until at length it threatened to extend its wanderings to the sea, through the hollow where the main part of Hilo lies. There seems to have been a strange fascination about this river of fire. People would go night after night from Hilo to watch it eating its way through forest and marsh, and sugar plantation, marking its passage by the smoke of the vegetation, and steam of the pools and streams it licked up as it crawled along. Their imaginations are enriched with stories of hair-breadth escapes from venturing upon the yielding surface, or from being caught in fogs in impassable hollows and ravines, and not knowing which way to turn, whilst the hiss, roar, and crackle of the advancing lava sounding nearer and nearer. The feelings of the population were raised to a terrible pitch at last, for the stream, though stopped by a slight rising at the bottom of the long ravine (whose mouth forms Hilo Bay), not more than a couple of miles above the town, was known to be massing deeper and deeper as if for a final destructive rush. On the right side it did overflow, and sent off a small stream as if to destroy a sugar-mill which the English vice-consul had just put up at great cost. His account of how they kept men always on the watch, and gangs of labourers ever ready at a moment's notice to put the machinery that could be removed on board ship, gave us some little insight into the anxiety with which the whole town was preparing for the exodus, into which it seemed only too probable that it would be forced. At this point, however, the flow hardened up for good, and the small lateral stream stopped within a few hundred yards of the sugar-mill.

29th Oct. Half-past seven A.M. found our party for the volcano landed on the bank of the small river, and soon settled into the saddle. What with Major and Miss Wodehouse, a couple of blue-jackets, two servants, and a man said to be a good blacksmith, who came from the sugar-mill, we made up a party of sixteen. The sugar-mill employés, who attended with the horses, talked pleasantly of the prospect of fine weather, but there was a dark bank of mist brooding over the horizon seawards, that looked suspicious. The first two miles of the road is good, and away went the party at a tearing pace, but the next two miles was less satisfactory, consisting of a bed of loose volcanic sand concealing numerous boulders, some round, some jagged, but all treacherously unsteady, so that we soon dropped from our dashing gallop into a sort of promiscuous flounder. After leaving the loose sand and boulders we entered on three miles of smooth rocks, interspersed here and there with natural flights of stone staircases, hedged in on each side with the huge palms, pandamus, and tree-ferns of a dense forest.

The sunlight streaming through the foliage made beautiful effects of light and colour. At the point of exit from the forest, our leader, Major Wodehouse, halted us for a few moments to rest the horses, and we then got on to three miles of swamp, like the fine sand, pervaded with boulders. Rain had set in and was falling steadily. After leaving the bog, the path sloped over a series of patches of hard lava, often twisted and twirled in a most extraordinary way. Four miles of this country brought us to the Halfway House, and we were glad enough to pile our dripping saddles, bridles, and cloaks in the verandah of the shanty, and prepare for lunch; but it was one thing to have lunch in one's mind's eye, quite another matter to get it into one's mouth. A case we had with us on being opened was found to contain six bottles of brandy, and a packet of sandwiches reduced to a pulp by the rain. The mules carrying provisions, alas! were miles behind. Things looked black, faces looked decidedly black, when one of our blue-jackets, who had been foraging in a native house hard by, returned with a bag of biscuits and a tin of beef. Manners and tempers softened at the sight, and the mules turning up much quicker than expected, we all found we were enjoying ourselves immensely. The young fellow who had come with us from the sugar-mill turned out a most useful and willing companion: his story was a curious one, for although of good birth and education, he had been left to his own resources from the age of fourteen. Taking employment in an iron foundry in Honolulu he soon became a good practical machinist, and thus secured his place in the sugar-mill, where he has a promising future.

On leaving the Halfway House the path was, if anything, more rugged and dreary than before, and we found the

process of ascending 4,000 feet under a dripping sky and over a series of lava terraces, decidedly monotonous. At length we arrived at a post which said, "Volcano House, one mile," the path here softened down into a kind of green drive, and we managed to shake our horses into a scrambling gallop, which brought us in a few minutes to an open grassy space, misty with driving rain and sweeping smoke, on which stood a large wooden framehouse with an inclosed paddock. In the hotel was a large central room with a huge fire-place, and very soon we found ourselves describing a comfortable circle round a pile of blazing wood. A treasure was unearthed in the shape of two big visitors'-books, with contributions from all sorts of persons. Miss Gordon Cumming had not only left a description, but had illustrated it with vivid effect. An English naval officer had filled a page with elaborate barometric registers; and the scientific staff of the *Challenger* had dropped a few sparks from the light of their intellects. Many enthusiastic persons had thought it well to leave a record in glowing epithets, which Mark Twain had most amusingly parodied, and several pages had effective caricatures. The Volcano House professes teetotal principles, but there was an abundant supply of most excellent milk, a thing that at sea one learns to regard as a most delightful luxury, and the call was so incessant that at last our landlord "guessed he'd have to meet so considerable a swallow by fixing a gallon or two in bed-room washing-jugs." The mist had begun to clear away and our eyes were perpetually being attracted to the window by outbursts of glowing steam, to all appearance but a very short distance in the hollow below us. After the meal was over we went out into the verandah,
30th Oct. but it was too wet to tempt us further. The morning found

the weather rapidly clearing, and we decided to have the guide in the verandah by half-past nine, as by that time the day would be either made or marred. To our great delight it turned out fine, and we started over the green lawn leading to the edge of the crater with a clear sky and bright sun over us, and just enough breeze to give a brisk feeling. We all had alpenstocks, though as far as the difficulties of the descent were concerned, they were an imposture, but the guide gave us to understand on the evidence of some of them having charred ends, that the lava flow required frequent testing, lest it should give way treacherously, and deposit one in the fiery flood surging beneath. For two-thirds of the way down, the path twists round trees and shrubs, and large black boulders, the steeper flights being relieved with steps made of logs of wood. The last part runs down a bank like a coal-pit, of loose volcanic ash. At the top of this was a white cross, marking the spot where a Mr. Holden had succumbed to heart disease; Major Wodehouse having been with him on his fatal trip. The hard lava, as a rule, lay in hummocks, much, one might imagine, like the ice in the Arctic regions that Captain Markham had described to us when we met him at Coquimbo, and when it caught the sunlight it gleamed as partially thawed ice will gleam. At first, especially where there were cracks and fissures in the surface, above which the hot air from below was quivering, one experienced some thrills of misgivings, but we derived comfort from the guide, who seemed anything but a bold adventurer, and we soon were convinced that he at any rate had no mind to be grilled. Across the floor to the edge of the lake is about three miles, but in the clear atmosphere it looks but a stone's throw. From all accounts the subterranean fire continually changes its points of energy. Major Wodehouse, who has paid several visits to Kilaua, described, as we walked along, that when he came with Captain Cator and Mr. Autridge of the *Scout*, small cones were slowly bubbling up all over the place, and that one which they watched, suddenly parting asunder up one of its sides, displayed a centre of glowing liquid of a sticky gum-like consistency. Sometimes it is possible to get within safe distance of the lava in this condition, and then persons incrust the halves of coins, or twist it round and round with their sticks into the form of a rude kind of vase. Quite lately two fresh lakes have developed themselves, and the broken mounds which rise to the height of at least a hundred feet above the oldest lake—the landlord at the Volcano House told us—are the work of not more than eighteen months. The general rule seems to be that when there is great activity about the lakes, the rest of the crater bottom is undisturbed, and that when the lakes are crusted over, upheavals of lava may be expected anywhere. At the moment we were there, the old lake, so our guide declared, was inaccessible, and it certainly looked so uncomfortably smoky, that we were quite resigned to turn away from its direction, up the gentle slope leading to the lake which has broken into existence last but one. In the crevices one's eye was constantly being caught by the golden glitter of the fine thread known as Pele's hair, which is formed by the wind catching the lava, when it is tossed into the air, and carrying off minute quantities, which harden into silk-like threads of a golden brown. The previous night, from the Volcano House verandah, we had watched globes of liquid fire driven out of the main fountain jets and glowing on the rocks, where they had been deposited by a

THE OLD CRATER, KILAUA.

lively breeze that was blowing, helping to form a further development of Pele's hair. The lake to whose brink we came in a few minutes, is roughly circular and supposed to be about a mile round. A bank of only a few feet high incloses the liquid—then there is a level platform, and outside this at varying distances, the main bank rises into a precipice—always steep, sometimes overhanging—of about a hundred feet. The surface of the hard lava within fifty yards of this is seamed with cracks, smoking with noisome vapours, and suggesting the possibility of large blocks falling forward into the furnace below. A rude parapet of lumps of rock stops the path. In eleven places round the edge of lake, spluttering columns of blood-red liquid were being tossed up with a sound as of distant muttering thunder. The rest of the surface was more or less covered with a black solid, or glutinous sheet, which here and there would bubble, and then shoot out a long crack, and gradually widen, until with increasing quickness the whole congealed film between it and the shore would roll itself up and be merged into molten tossing fiery blood. The immediate brink of the lake had a very curious look, being furred over with the golden-brown Pele's hair. Pritchett made an excursion to the highest point of the bank, but any picture must of necessity very inadequately express the marvels of this sight, for the effect is dependent on the incessant change of action. We tried to poke our sticks down the fissures close to us, but did not succeed in scorching them. Sitting down, however, with one's legs over a crack, we became very successful in obtaining evidence of the earth's internal heat, and felt very like a cook who had mistaken the range for one of the kitchen chairs. After looking at the lake for about an hour we started back for the hotel, and after lunch started off in search of a most perfect specimen of an extinct crater, reported as being stowed away in the forest a mile or two off. We found it to be perfectly circular, about 250 yards in diameter, a thousand feet deep, with sides clothed with scrub and stunted trees. On looking down we thought the depth of a thousand feet must be exaggerated, but one of the blue-jackets who climbed, certainly not more than half way, with an aneroid instrument, confirmed this measurement. Resting on the turf under the trees by the edge, a kind of owl paid us a visit, almost the only wild bird we saw on Hawaii, where it is said all the birds eggs are eaten by rats.

We hurried over our dinner at the hotel, as the moon was at about its second quarter, to make a moonlight circuit of the great crater's brink, to the point where, some two and a half miles distant, a rising bluff was reported as overlooking the old original lake. It was hardly worth the trouble of getting saturated with dew; certainly we gained a view of the latest and third point of volcanic outbreak, where there was great show of fiery activity, as if the lake was then and there extending its boundaries, and inclosing large portions of the surrounding lava field. Just before going to bed we got a beautiful picture. The moon stood almost vertically above the fire-lakes, which by the intensity of their heat expanded and drove upward a column of air drifting above them, mingling it at the same time with some of their own steam, and the top of the column condensed high up in the cooler sky into a mass of soft cloud, with its lower half rose-red from the lava fires below, its upper surface snowy white
31st Oct. and gleaming in the downpour of moonlight. The morning was again windy and rainy, and we were all rather glad to

find ourselves, after the long tedious ride, safe and sound again, and sitting down to a comfortable dinner on board the *Wanderer* at seven o'clock. When we got on board we found the men had caught a patriarchal shark which one of the natives professed to recognise and describe as a well-known character known as "Hilo Tom." When he was cut up, they tested the size of his jaws, and found the biggest man in the ship could pass easily between the rows of his jagged and well worn teeth.

## CHAPTER XVIII.

*QUEEN EMMA AND QUEEN KAPIOLANI—MAUI—FROM HONOLULU TO YOKOHAMA.*

1st Nov. AT 8 A.M. we left Hilo with a fresh north-east trade in our favour, a bright lovely day, followed by an equally
2nd Nov. beautiful night with a clear moon. At 4 A.M. we were close up to Honolulu, lying to and waiting for daylight; by eight o'clock we were again moored in the harbour, and after breakfast parted with great regret from our visitors, Major and Miss Wodehouse. The Russian admiral and his staff came on board to return the call made on him at Hilo, and in the afternoon, by his invitation, we went on board his flag-ship the *Africa*, a very fine, roomy, and heavily armoured vessel, having all the latest inventions in Krupp and Nordenfelt guns, and torpedoes, but in her construction she did not seem to be much stronger than a merchant steamer, as she had very thin sides and no armour, and the upper part of her compound engines stood some three or four feet higher than the upper deck, and were quite unprotected. The officers' accommodation is very good, and the admiral's cabin, with its panelling of coloured woods, excited much admiration. In the evening there was a ball given by the king (who had returned to Honolulu during our absence) to the Russian officers, to which we were all invited, but only

three of the gentlemen went. In the afternoon of the fourth, we drove out to Waikiki, to call on Princess Likiliki, and heard that the king had enjoyed his visit to England very much. Next day Queen Emma and her ladies in waiting, with Major, Mrs. and Miss Wodehouse came off to lunch with us, and on the 7th we went to the Roman Catholic church, to hear a *Te Deum* for the king's safe return, and had capital seats in the French minister's pew. Princess Ruth looked gorgeous in a dress of black and gold satin.

3rd, 4th Nov.

5th Nov.

6th Nov.

7th Nov.

HONOLULU.

The band played at the king's entrance and as he left the cathedral. After this we had plenty to do in writing invitations for a dance we proposed to give on the 11th. Queen Emma, who was going to start for the Island of Maui in a small steamer, consented to put off her departure until the 14th, and let us take her there in the *Wanderer*.

To-day we had the pleasure of entertaining Mr. Cleghorn, Princess Likiliki, and little Kauilani at lunch, and in the evening went to a most delightful dance given by Mrs.

8th Nov.

Wodehouse, where we had another opportunity of admiring the beauties of Honolulu.

9th Nov. Next day the king and his chamberlain, Colonel Judd, came to lunch, the king was full of conversation about his visit to England, and expressed himself as delighted with his reception by the Queen, and Prince and Princess of Wales, saying that they were "just as kind as they could be."

10th Nov. On the 10th we had another pic-nic to the Pearl River,
11th Nov. and on the 11th our dance came off. It was a dull damp morning, and looked as if likely to rain all day, but about 4 P.M. it began to clear up. We had some of the town band to play, and danced in the saloon. About 9 P.M. some fireworks we had brought from England were sent up from a boat, and we did not lose our visitors until past twelve at night, and I think all went off very well.

12th Nov. The next morning little Kauilani came off, bringing Mrs. Lambert many pretty presents sent by her mother; the little "Hope of Hawaii" had got to feel quite at home on board, and played about quite merrily. Later on, Mrs. Lambert and I went ashore to lunch with the King at his country house at Waikiki, and were presented to the Queen Kapiolani; she is tall and handsome with most pleasant manners, but unfortunately for us, did not speak English. The King, who took Mrs. Lambert in, was most chatty and kind, and we had an excellent lunch, with everything done in capital style. About three o'clock we returned to town in three carriages with the King and Queen, who took us to the palace to show us the beautiful feather mantles, cloaks, and helmets, most interesting relics of past days, and very valuable from the scarcity of the royal feathers. We then took our leave, and felt inclined to say of this royal couple, what the King had said of our Prince and Princess, namely, that they were "just as kind as they could be." The next day, Sunday, several friends came off to service, 13th Nov. and in the afternoon the King and Queen paid us a visit, bringing a lady interpreter for the Queen. We were all charmed with her majesty's pleasant, courteous manners, and the kind interest she showed in all there was to see on board ship. This was a busy day on board, all hands busily 14th Nov. employed washing the ship, and hoisting in live stock and other stores for the somewhat lengthy voyage before us, to Yokohama. We all went ashore to a farewell lunch at the Wodehouse's, and afterwards to see Princess Ruth's beautiful new house, which she showed us over herself. The reception rooms are panelled with the woods of the various islands, and are lofty and handsome; after saying good-bye to this kindly lady, who loaded us with presents, we parted reluctantly from Major and Mrs. Wodehouse and their charming children, of whom we had all grown very fond, and went on board again to receive Queen Emma and her suite. We found many pretty presents from both the King and Queen, and from other kind friends, a number of whom had come off to wish us God speed on our voyage. It was with feelings of real regret that we left the hospitable, generous people with whom in five weeks we had formed friendships as intimate as if they had lasted for years instead of days; and Honolulu, with the recollections of the court and royal personages, the Wodehouses, and other friends, American, French, and English, will always have a very warm corner in our hearts. At 5 P.M., Queen Emma and her attendants, about twenty in all being on board, we started on our voyage and a rough unpleasant one it proved, for when we turned the island of Molokai, it was to encounter a stiff north-east trade and heavy head-sea making us pitch violently; however, there was nothing to be done but peg away into it, our guests alas disappearing in various conditions of sea-sickness, with the exception of Queen Emma, who proved an excellent sailor. At 10.45 next morning we were at anchor in the harbour of Maui. A reception committee very soon came off (having seen the Hawaiian ensign at our main-mast head) to welcome the Queen, and tell her of the preparations made for her reception. From the deck we could see crowds of people on horseback, riding along at full gallop to the landing place, and who there formed a most enthusiastic and delighted mass. It had been our intention to land the Queen and put to sea at once, but both her majesty and the members of the committee so pressingly insisted upon our going on shore on the morrow to witness her drawing-room and see her people make their offerings, that we had no choice in the matter. When I took the Queen on shore it was a most touching sight to see the people all kneeling round her pressing forward to kiss her dress, and it was with difficulty I placed her in the carriage, and returned to the ship, from whence we watched the huge procession of carriages, horses, and foot people wending its way along the sea shore to Wailuku where Queen Emma was to reside. At 9 A.M. next morning our whole party landed and got into three carriages sent for us by the reception committee, four of whom accompanied us on horseback. It was a drive of about two miles, but some time before we arrived at the house we overtook long lines of men, women, and children, each carrying in their hands some offering for the Queen, of pigs, fowls, ducks, turkeys, fruits. When we arrived at our

15th Nov.

16th Nov.

destination we found the Queen standing under a large bread-fruit tree in front of an arm-chair, and on a handsome mat. She was surrounded by attendants, and her chamberlain held over her head a staff ornamented with the royal yellow feathers. By the side of her chair on each side were placed others for our use. Having received us most kindly, the band played "God save the Queen;" we then sat down, and the Queen received homage and tribute from all, the tribute being placed in the hands of the attendants. It was really a most affecting sight—the deep love and reverence that the people showed for their Queen. Each comer in succession knelt down upon the farthest edge of the mat upon which she stood, about six feet off her, and then crept up to her, with earnest fixed gaze, many in tears, and so, kissing her hands, and skirts of her dress, made way for the next. The numbers seemed as if they never would come to an end, for when we left about noon, long strings of people still lined the road side. We were told that we must not go away without having an interview first with the reception committee, who said it was the wish of the people, as a token of gratitude to us for bringing their dear Queen, to present us with some offerings that might be of use on our voyage to Japan, and that as they could not place them before us at that spot, I was asked to go a little way with them, that they might show me the present. I accordingly walked a little way with these good people, and found to my astonishment that the offering consisted of no less than four large cartloads of every kind of fruit and vegetable, turkeys, ducks, fowls, and pigs. I assured the committee, that while I fully felt their kindness, I could not take the things, for I actually had not got room for them on board my ship. It

was evident, however, that they would be so hurt at a refusal, that as a compromise I said I would take as much as I could make room for, which consisted of about half a cartload; and told them that my bringing Queen Emma was an honour to me, as she had intrusted herself to my care, and that she was known and loved by our Queen, who loved all good and virtuous women, such as their own Queen most certainly was; I hope they forgave us for leaving all the wealth of produce that they had collected; and so we bade adieu to our kindly entertainers, and hiring a boat to take off our present to the ship, started at 3 P.M. for our voyage to Yokohama, feeling more than ever the excessive kindness that we had received at all hands during our stay
17th Nov. in Hawaii. With smooth water and a long day we passed round Molokai Island again in the early morning, and at noon we had run 130 miles. We were now really Westward
18th Nov. Ho again, and homeward bound. Still fine clear weather, but only easy steam to help us along up till now, when a nice
19th Nov. north-east breeze enabled us to feather the screw, and under square- and studding-sails, we ran along at a pace of between
20th Nov. five and six miles an hour. We had service as usual, and passed a very pleasant quiet day, running with the same
21st Nov. to 5th Dec. canvas as yesterday, fine weather continuing. Our voyage to Japan was indeed singularly uneventful. Those who are interested in such matters will be able to trace the progress
5th Dec. we made by reference to the Appendix. On the fifth we had done with the trade-wind, and changed our course to the northward, taking in the square-sails and going on under steam again, and with fore- and aft-sails.

6th Dec. To-day we had calm weather, and at early morning saw a beautiful eclipse of the moon, lasting from 1 A.M. to 5 A.M.

in a cloudless sky. From to-day until the tenth we got tolerably fair weather, but on the last of these days the wind headed us, freshening with squalls. At 3.5 P.M. we passed the island of Fatsizio, and on the following day sighted Nosima-saki light at 1.40 A.M., and lay to for daylight, standing in for land at 5.20 A.M., steaming through an immense fleet of fishing junks, which with the pretty wooded hills and islands made a very pleasant and cheerful picture. The first island we passed was Vries, which is nothing more than an active volcano, said to be a sort of safety-valve to the larger but extinct Fusiyama. As we passed, it was emitting spasmodic volumes of white smoke. Before casting anchor, Captain Pollard of H.M.S. *Zephyr* came on board, and kindly pointed us out a good berth, and spent the afternoon. The little *Zephyr* looked very familiar, for during her previous commission she had been commanded by our son-in-law, Captain Clutterbuck on the North American station, and it seemed but a few weeks ago that we had gone down to Sheerness to see him and his ship previous to her being paid off. We did not in fact want for visitors, being soon boarded by officers from Japanese, French, American, Russian, and Italian men-of-war, all offering us their kind services. It proved a lovely afternoon, and we thoroughly enjoyed the novel scene around us, though the cold after our long stay in warm latitudes was rather trying. The enjoyment of the day was completed by the arrival of a welcome batch of letters and newspapers.

7th, 8th, 9th, 10th Dec.

11th Dec.

The ship to-day presented the appearance of a fair; it was impossible to keep the people off, who brought curios of every kind for sale, boxes and cabinets, brooches, &c. At two o'clock we went ashore, and enjoyed our first ride in jinrickshas.

12th Dec.

These vehicles are like large perambulators drawn by one man. Off we went, a party of ten, all in fits of laughter; the men who drew us apparently as much amused as we were, running along at a good pace, and occasionally giving a skip like children playing at horses. We only took a passing look at the beautiful things we saw—bronzes, lacquer, china, silks, embroidered dresses, and having called on our consul, trotted back to the landing-place at half-past four. We were guided by Mr. Welsh, whose services we were lucky in securing during our stay in Yokohama, Yeddo and the neighbourhood; he is a most excellent guide and adviser, having a good knowledge of the Japanese language, and being very active, obliging, and trustworthy.

13th Dec. Mr. Welsh started off to see if he could secure rooms for our large party at Yeddo, and came back in the evening with the good news that he had been able to do so. During the day we went ashore, and stared about at the people and the shops—but the town has been so well described by Miss Bird lately, and many other writers, that I will not weary my readers with fuller descriptions. In the evening we went to the Gaiety Theatre, to see a company of amateurs perform the *Palace of Truth*, which was exceedingly well done, the ladies' parts charmingly played, and the dresses perfectly magnificent.

14th Dec. Our whole party, including the servants, left the ship in time to get off by the 9.15 A.M. train for Tokio (Yeddo). The whole country on either side of the line was cultivated in small patches as elaborately as any English nursery garden. Approaching the town Mr. Welsh pointed out to us the scene of Sir Harry Parkes' gallant resistance to a murderous attack made upon him and his guard by a jealous Daimio. Getting into the jinrickshas at the station, a passing foreboding entered the mind of certainly one of our party, as the eager eye of the man was marked, who was preparing to wheel one along in his "one-man-power vehicle," for a French newspaper lying in the waiting room at the Yokohama station contained a paragraph, telling of a compound fracture sustained by a doctor of a French man-of-war by an upset of one of these go-carts. We found the Seiyoken hotel an excellent establishment, although there was not a single woman on the premises, and beautifully clean and comfortable. There is a French cook, and when you find you can revel in his dishes, and are lodged for three "*yen*," that is, paper dollars a day, one wonders that people in search of cheap living do not come out to these islands. Shortly after lunch we were trotted out in fourteen jinrickshas to see what was described as "the high class music and dancing."

Two or three miles ride through the streets of Tokio brought us to the doorway of the house, where we gentlemen had to take our boots off before entering, and not without reason, for the walls, floors, chairs, in fact everything in the house was perfectly clean and spotless. A large room on the first floor, with sliding walls, and paper windows, and a sunny balcony overlooking a sacred piece of water, was the scene of the "high class" entertainment. Heavily dressed, with hair a composition of gum, combs, and amber pins, the girls postured rather than danced on soft mats, working their fans and striking their quaint attitudes with a faultless precision, telling of years of that extraordinarily patient care, found in everything connected with old Japan; others accompanied them with voices which made you wince with their shrill discord, and the twanging of untuneful instruments. Coming out, we saw jets of smoke issuing from the roof of a house opposite. A fire seems to amuse the Japanese rather than to cause serious distress, for Mr. Welsh told us that the other evening he had seen a woman sitting in the street with her children, laughing at the freaks of the flames as they licked up her house. "Why do you laugh?" said he. "Why not?" she answered. "I cannot cry my house out, so I may as well get some amusement out of it." After dinner we went into the main street to try and pick up curiosities from the stalls along the side paths. Wetherall succeeded in securing a bronze tray with silver mountings. Somebody said it must have been very old, no doubt out of some Daimio's house—it was known many of them were selling their treasures—some people always are in luck! By the light of the hotel lamp, however, it was seen that the rare antique bore a circular stamp, "Muntz's patent metal." The secretary of the Mikado's household called when we were out, and was kind enough to say he would come again on the morrow.

15th Dec. We started early next day to see the temple of Sheba, magnificent in its proportions and decorations. It stands in a beautifully kept park, but has been well described both by Lady Brassey and Miss Bird. We had lunch in a restaurant in the park, from which we got a lovely view of Yeddo. This place is very typical of the present transition state of Japan, from mediævalism to modern civilisation. On a platform above the restaurant was an image of Diabutz, some twenty feet high: below, an old bell-tower boomed forth the hours of prayer to an inattentive district. Many feet lower down lay a sacred lake, brown with winter rushes, whose waters coldly washed the sides of a holy island; whilst inside the restaurant, we sat regaled with elaborate French cookery. As we were at lunch, in came Dr. Gray and Mr. Tyacke, bringing a budget of English letters. After these had been enjoyed, we began a further series of temples, tombs, and industrial museums, and some beautifully kept gardens with a cage of bears. The courts of the last temple are a sort of blending of a bazaar and pleasure fair. Within the space of a hundred yards we bargained for a bronze bowl, bought a pipe ornament, watched a man mutter his worship to Buddha, looked on at a conjuring performance, and strolled through a waxwork exhibition, representing the adventures of a pilgrim and the perils he escaped. In the evening we dined at the Legation with Mr. and Mrs. Kennedy, and after a very pleasant evening, returned to the hotel to sleep.

This morning the two boys, Pritchett, Wetherall and 16th Dec.
I started between 6 and 7 A.M. by the invitation of Mr. Nagasaki, the king's chamberlain, to enjoy a novel sort of duck catching in the park of one of the Mikado's summer palaces. Some drove and some walked until we arrived on the ground, each of us being armed with a long-handled net not unlike a butterfly net, but much stronger, the handles being made of tough but light bamboo. We made our way by a one-span bridge into the grounds, ornamented with clipped firs and cedars, and artificial pools. Then through an old tilt-yard with high turf banks and ground of hard sand, until we came to a small hut, with a kind of sunken trough in and along the centre of its length, in which were placed Japanese braziers containing charcoal fires; here on benches on either side, with toes to the fire, pipes in mouth, we waited until a keeper stole up to the door, and bowing almost to the ground, announced in a whisper that a decoy was full. Silently we crept over the turf after our guide, who

assigned to each his position by the side of a deep ditch, with a turf rampart about two feet six inches high to keep the sportsman out of sight of the ducks, which are scared by the least sound. All being placed, a keeper stationed at the extremity of the line discovered himself, and immediately afterwards up rose the wild fowl. We managed to bag three teal, a fourth got away, but a falcon was slipped, and before the teal could clear a high clump of bamboos in the park, he had seized him in his clutches. This decoy being empty, the same proceedings were gone through with others as they filled, until about 11 A.M., when we sat down to an excellent breakfast in a pretty little summer house. Presently a flutter amongst several of the attendants grouped about drew our attention in their direction, and we saw them exchanging bows with a gentleman coming towards us, decorated with many orders. "The Prince Imperial!" exclaimed Mr. Nagasaki, who presented us in due order. The prince took up a net and led us to fresh onslaughts upon the duck. Max's sporting instincts were stronger than his respect for royalty, and he not only bagged the prince's duck, but almost caught the prince's head in his eagerness to bring his net smartly down upon the rising bird. However, Prince Higashi Fushimi only smiled very pleasantly, and invited us all to a most splendid lunch, at which the ladies of our party joined us, and where we met several members of the English Legation, Mr. and Mrs. Kennedy, and some of the Mikado's ministers. It was given at one of the royal residences, situated in lovely grounds, with a lake in front, everything being done in European fashion. The prince is a very pleasant and courteous gentleman, and speaks with great pleasure of his visit to England, and of the kindness shown him by the

Prince of Wales. After lunch we went off in jinrickshas to a kind of government bazaar. The prices seemed absurdly small, and some of the party expressed the feeling that one would really save money by making purchases. Exquisite little cups, Satsuma ware, lacquer, combs of ivory and tortoise-shell, bronzes, carved wood, and other innumerable articles of beauty, were piled up in the most tempting fashion. After dinner Welsh gave us an account of the Japanese theatre, and of plays which went on all day and most of the night, one play taking a week or a fortnight. In the course of conversation it turned out that the Tokio theatre was the largest and best in Japan, and that a star of special brilliancy was now performing, so at half past-eight three or four of the gentlemen turned out into the dark street. There was a little delay in getting jinrickshas, as the men were warming themselves over the braziers in the tea houses. The main artery of Tokio is lit with gas, so that as long as we kept to that, things wore a comparatively homelike look, but when we rattled round a sharp corner into a smaller street, we found ourselves among fantastically shaped, and coloured lanterns, and realised the strangeness of our surroundings. The perfect good temper of the crowd was most striking, the Japanese equivalent of "get out of the road" was perpetually dinned into the ears of the passers-by by the men who drew us, and instead of a scowl or oath, evoked only a laugh, although often followed by a downright hustling from the wheels. Presently a street all lanterns, tea houses with open fronts, and crowds of jinrickshas announced the presence of the theatre. A man at the door told us we were in time for the twenty-fifth act, but he added "the theatre would soon close for the night." Finding, however, that "soon" meant an hour, we went in, and passing the refreshment stalls scattered about the floor in front of the building, mounted a very steep flight of stairs into a long passage, on one side of which were small apartments, and on the other the *boxes* —square pens, bare of furniture, separated from each other by a boarding about a couple of feet high, and just large enough for the four of us to find squatting room. The floor of the house was divided in a similar way, but as many as eight or ten Japanese found stowage room in one of these. The stage was long but narrow. At each end was a raised part where the musicians sat, five on each side, with drums and Japanese guitars. The action of the play seemed to be both melodramatic and farcical. A family of marine divinities were sitting in state at the bottom of the sea, which was represented by a paper background, coloured with innumerable, curling, green waves. After going through much elaborate posturing and talking, the chief personage retired through the background of sea, the curling billows opening for him to pass through. When he had disappeared there was a tremendous outburst from the orchestra, and an adventurous lover of the goddess descended from the ceiling disguised as a modern diver. Then came some pantomime business, the diver kicking about, suspended in the air, and set upon by the attendants. After a struggle his air-pipe was cut away, and he was apparently first drowned and then brought to life again. The goddess then came to the front and performed a sort of amatory minuet with the diver, until the reappearance of the husband through the waves caused a break in the dance. Appearances were, however, saved by the adroitness of one of the female attendants, who substituted herself for her mistress, and received the diver's attentions. The scene closed with a breakdown by a couple of men, who danced off to a hideous rattle of drums, amid which, scenery, orchestra, and all the stage properties were hurried away, almost before we could think of getting our coats and mufflers to face the cold night air outside.

JINRICKSHA.

## CHAPTER XIX.

*KOBE.—OSAKA.—KIOTO.—HOLY ISLAND, AND THE INLAND SEA.*

17th Dec. FLUTTERING snowflakes, white roofs, and ground half snow half slush, greeted us on the next morning, and made us, after a morning of shivering about the hotel, anxious to get back to our quarters on board the *Wanderer*. Some younger members of the party found an occupation in getting tattooed. This is forbidden by an edict of the Mikado, but the former prevalence of the practice has called into existence an artist who is so talented that he is greatly patronised by foreigners. His book of patterns, dragons, birds, reptiles, and mythological beings, was enough to charm any one into the folly of carrying off a Japanese curiosity attached to his skin. Our exodus from the hotel was through driving sleet, but the discomfort was somewhat relieved by the grotesque appearance of the jinricksha men, who, thatched with straw, reminded one of ambulatory champagne bottles. At the station, too, we fell in with some Coreans, part of an embassy sent to Tokio to settle differences, and discuss appearances or Russian aggression in their part of Asia. They are chiefly remarkable for their broad-brimmed hats, with open horse-hair work in the upper part of the crown. One of them, who looked about twenty, had a singularly good-tempered

TOKIO.

and lively look, and laughed with every muscle of his body, as he looked at us with a critical eye, and said inquiringly: "Yankee? English? French?" On arrival at the landing-place we found the contractor's steam launch with a covered cabin, which put us on board warm and dry.

The storm of yesterday had cleared the atmosphere, and Fusiyama and the mountain ridge shone out sparkling white, and clear cut in the bright sunshine. Welch had brought us off several great braziers and a supply of charcoal, so that we made ourselves very comfortable and defied the cold. 18th Dec.

Another bright cold day. The sampans brought the same stream of curiosity dealers as had invaded us on arrival. Two antique gods carefully carved were offered for eight yen. *Four* were proffered, and accepted with so sweet a smile and gift of a mirror as "cumshaw" as the sailors call what is I believe "komisha" in Japanese, or "something over," that the purchaser regretted he had not offered *two*. The tattoo artist came from Tokio, and his skill was marvellous. One of the gentlemen of our party selected from the book of patterns a funny figure that looked like a caricature of a native water-carrier. Half-a-dozen turns of the artist's wrist, and the outline in Indian ink was perfect; then with wonderful rapidity, now using a holder with a single needle, now one with several, and only drawing blood where the shadows were at their deepest, he worked into the skin in blue and vermilion all the elaborate details, hair, toes, fingers, even the stitches in the great puffed-out skin on the figure's back. The patient declared the pain was only of a slight and worrying nature, but after about three quarters of an hour, he appeared very anxious for the artist not to exert himself 19th Dec.

for too long at a time, and proposed that he should regale himself with a pipe. The suggestion was readily accepted, a little brass implement was pulled out with a bowl about the size of a pea, which was filled, lit, and puffed out in about two minutes, and the prodding began again with fresh vigour, to the disappointment of the patient, who had thought he might get a quarter of an hour's respite; when it was all over the sufferer declared that he experienced an *agreeable* glow all over his body. We had Prince Higashi Fushimi, the Minister of Marine, and Mr. Nagasaki to lunch, and Mrs. Kennedy kindly sent us a Japanese lady artist, who gave us specimens of her skill in painting on silk and paper for house decorations.

20th Dec. Mr. and Mrs. Kennedy, and a large party from the Legation, lunched and spent some time on board, after which our time was taken up with making our final purchases. In the evening we had some friends to dinner, amongst whom was Mr. Layard, lately attached to the Legation.

21st Dec. Continued bright and still weather. We left Yokohama at 8 A.M. In making a detour through the shipping to exchange cheers with the *Zephyr*, a sampan ran foul of us, and her mast went by the board. As we rushed down the long gulf, Fusiyama's white cone glittered in the bright sunlight on our starboard side, and on the port side Vries poured out volumes of smoke and fine ash. The sea was alive with porpoises, and ruffled with a shoal of smaller fish, over which innumerable sea-birds fluttered, swooped and wrangled.

22nd Dec. We had a fine clear night, with light breezes; at 1 P.M. we passed between the mainland and the island of Ooshima, a lovely little strait full of vessels of every nationality, lying in shelter in anticipation of a blow from the S.W., which the glass

KAWACHI, INLAND SEA, JAPAN.

indicated. Emerging from the strait we were greeted by a wild gust of wind and a rainy gale dead in our teeth; fortunately a bend in the coast soon brought the wind first abeam, and then on our quarter, so that we went up the gulf leading into the Inland Sea at a spanking pace, and at 4 A.M. 23rd Dec. we were in Isumi Strait, with Tomengai Sima light on our beam, and anchored in Kobe roads at 7.45 A.M. We found the *Encounter* and *Flying-fish* lying here, as well as some large steamers, and many fishing boats. It is a pretty land-locked harbour, with a straggling town and wood-clad hills. Captain Robinson of the *Encounter* came to lunch with us, and in the afternoon we went ashore and found plenty of amusement in looking at the shops and curiosities. In the evening we sat in the poop cabin, with lamps lighted and good fires in the braziers, enjoying a musical evening.

We had heard that Kobe was free from very wintry 24th Dec. weather, but a sharp wind this morning suggested the probability of snow; the thermometer too was at thirty-four. About the middle of the day a dark squall swept down the mountain behind the town, flinging showers of white flakes as it sped along. A white spot near the top of the ridge, which we had put down as a patch of snow, turned out to be a Temple of the Moon, and we decided to visit it in an afternoon walk. The level ground and slope between the coast-road and foot of the steep ridge showed an infinite ingenuity and patient labour of irrigation. Crossing this by an angular narrow pathway we came upon patches of cultivated ground, whose dark green had shone out conspicuously from the ship's deck; they turned out to be tea fields, the shrubs about eighteen inches in height. Every ten minutes or so we passed a shrine, with the stone lanterns still full of prayer pebbles, but we did not see any one add to the heap, and our guide smiled as we looked at them, with an air of conscious superiority. From the top, a fine view was got, but the temple was not worth the walk, after the fine ones we had seen at Yeddo.

25th Dec. It was well we made Kobe when we did, for the Yokohama steamer arriving this morning reported a frightful gale from which it had been obliged to shelter for nineteen hours in one of the coast harbours. The morning was very still. As it was Sunday our Christmas festivities were put off until to-morrow. The pilot we had on board was a very amusing guide, and full of experiences. He told us that when cruising off the Southern Island of the Marquesas, about twelve years ago, one of the crew of the ship he was in died, and that they put into a little bay that was sheltered from the run of the swell by a high promontory. "It was very awkward work," said he, "landing the body through the surf, and when we did get him ashore, we had to get the chief to taboo the grave, as the people were all standing round ready to dig him up and eat him as soon as the funeral was over. Another day when we were ashore we heard some screaming in a hut, and looking in found the people holding down a young girl, and tattooing her feet and legs with boots and stockings. Her hands had already been done with gloves." It was somewhat remarkable that this turned out to be the very Bon Repos Bay where we saw the burial places, and where one of the sights was a young woman tattooed in the way he described.

After service the ladies pulled round to wish the officers of the *Encounter* and *Flying Fish* a "Merry Christmas," and were obliged to go on board and have a glass of champagne, and look at the men's dinners. Our men had decorated the *Wanderer* with evergreens from stem to stern, as well as the figure-head—it had been done at night, so was a surprise to us all. The forecastle looked very cosy too with flags and Chinese lanterns.

Next day was kept as a holiday; and we had some native **26th Dec.** conjurors on board who performed capital tricks. The only drawback to the festivities was Barnes the carpenter falling down the forward hatchway and breaking his leg.

To-day all our party started off by train to Osaka, accom- **27th Dec.** panied by the Japanese official who travelled with us by the courtesy of the Japanese Government, who have shown us every kind of civility and kindness. The rail was a very narrow single line, with long saloon carriages, and traversed the tract of level ground skirting the ridge of mountains, and extending into a far-reaching river plain. The fields wore a wintry barrenness of aspect, their monotony only relieved by the corn strewn upon the ground and left until threshing time, and tree stems protected from the cold by swathings of straw, almost thick enough here and there to look like conical ricks. About an hour brought us to Osaka, and a whole host of third-class passengers poured out upon the platform, but very few first or second class passengers beside ourselves. We went through narrow streets, in our long train of jinrickshas, and drew up in front of a large inn kept by a native, but fitted up with an eye to foreign visitors, where we sat down to a capital luncheon with numbers of very good oysters. After lunch we took jinrickshas again to a huge relic of the feudal days, a cyclopean castle, the former centre of the Tycoon's power, where he ruled from an impregnable position, and overawed the Mikado—an enormous moat lined with giant masonry. wall within wall, built of cunningly fitted blocks of stone of all shapes. From the top of this we got a fine view of the town and surrounding district. Masses of temple towered over the low monotony of wooden houses. Far away through the clear atmosphere we could make out the ships of war lying off Kobe. Just outside the moat was a brand new gun factory, and on an open space on the other side were bodies of troops, drilled and armed after European fashion. After leaving the fortress we had only time to take a hurried glance at one of the old Shinto temples, where we found not only live but wooden horses. The Japanese attaché told us that the horses are not objects of worship, but merely sacred as being used on high days for the chief priests to ride on in procession. Getting into the train again we sped along until dusk, when we arrived at Kioto. The jinricksha men, who were harnessed tandem, showed the sort of eager skittishness that a young hunter does in the presence of hounds; like kittens and children they are charged with a large amount of superfluous go, and in addition to the pleasure of working this off, is the thought, no doubt, of the rice and the saki to be got by the extra fares that foreign visitors give. There was no gas to relieve the gloom of this city of old Japan, until lately unapproachable by foreigners, the sacred centre of the Mikado worship, where the Dutch envoys used to come annually and perform all manner of prostrations and servile antics for the amusement of the court. We must have rattled along for as much as three miles, when we crossed a long bridge over a half dried river bed, and began a tolerably steep ascent, which after about a quarter of a mile led us up to the doorway in the

high wall of the old Japanese nobleman's house, now doing duty as an inn. Steps communicated in rather a confused manner between the different parts. As at Osaka, there was a large room on the first floor for general eating purposes, which was intensely hot with a stove. The inn is under the same management as that at Osaka. Strolling about after dinner, notwithstanding the cold and occasional snow, the streets were crowded with chattering natives, who showed the same cheery good humour as noticed before. The bridges across the river are most suitable for tragedies—dark, and with low parapets; a man might be stabbed, and sent rolling over the pebbly bed with little prospect of discovery. Bedtime raised an unpleasant question as to whether it were better to freeze or to choke; the rooms left to themselves were bitterly cold, and the only means of warming them lay in having recourse to charcoal braziers; finally we decided to freeze. Breakfast was delayed the following morning by a perfect exhibition of specialties of Kioto industry. In the 28th Dec.
neighbourhood there is a mountain consisting almost entirely of porcelain clay and rich copper mines. Under the patronage of the Mikado's court, a special school of pottery and artistic bronze-work has arisen. Spread over the chairs and sofas were silks, quilted dresses, and hangings stiff with embroidered storks and imaginary animals, executed with wonderful vigour. Crowded over the tables were teapots, jugs, and all sorts of boxes and vessels of inlaid bronze. Our visits in the town were chiefly directed to the workshops, the most interesting being those for porcelain and metal enamel. Outside the enamel workshop, a variety of details coming together made a quaint picture. The street had been cut through a graveyard, old enough I suppose for the

identity of the buried corpses to have been lost sight of, for the roadway was paved with fragments of the lantern tombs. A pack bullock, with angry quivering nostril and flashing eye, was lashed by his nose-ring to the roof of a house hard by. Amongst the jinrickishas the men were bear-fighting, with the strangest horse-play. Having but a short time at our disposal we visited only one temple, that of Nishni Hongangi, which was on a stupendous scale, with its massive pillars and masonry, and huge brass bell ten feet in diameter at the mouth, and fifteen or twenty feet high, and with ornamental gardens and ponds cut and scooped out of the rocky hillside. The Japanese *attaché* introduced us to the chief priest, who conducted us to chairs of black wood very hard and heavy as iron, which stood in a magnificently hung and ornamented waiting-room, and insisted on regaling us with cakes and sweets, tea and Japanese wines, and then showed us all those parts of the temple not usually visited by strangers. Had he any historic instinct, one would think he must have contrasted the present with regret, when a party of foreign visitors were perhaps his most respectful pilgrims, and where the guides spoke in language of contemptuous patronage of his religion as a worn out institution only tolerated for the extreme ignorance still lingering amongst the dregs of the people. From the temple we paid a visit to the Kioto bazaar, which has the merit of having all the articles marked at a fair price, so that extortion is impossible, after which we took train again, arriving on board about 6.30 P.M. Mr. Simpson of the *Encounter* dined with us, and although we had had a very enjoyable trip we were not sorry to get back to our own beds again.

WATER GATE OF SACRED ISLAND, INLAND SEA, JAPAN

Next morning we left Kobe at 11 A.M., a bright but very cold morning, and steamed close to the *Flying Fish* and *Encounter* to say good-bye. The government official was to remain with us until we reached Nagasaki, the southernmost town of Japan. We stood through the narrow Akashino Seto channel into the Inland Sea, and anchored for the night, at 5 P.M., in the snug and pretty harbour of Sakati in the island of Sozu-sima, leaving at 6 A.M. and going on through narrows thickly studded with pretty islets, on each of which we could see small towns and villages. The hills on each side of us were terraced and planted to the tops, every inch of available ground being cultivated. Emerging from these straits, we crossed a wider expanse, and then entered the still narrower strait of Mekari Seto, passing many islands, and the mainland of Nipon. On coming to the outlet we cast anchor at 4.15 in the harbour of Mitarai, which is completely land-locked, and situated between the islands of Mitarai and Okamura.

29th Dec.

30th Dec.

We were off next morning at 7 A.M., and crossing Midima Nada we continued passing innumerable islands, boats, and junks, until we anchored in Miya Shima harbour, in Itsuku or Holy Island. The town is very small, and clusters round the temple, or temples, which are situated in a nook below the green hills. It is a most lovely spot, and visited by pilgrims from all parts of Japan, who land from their junks, and pass in a boat under a curious arch in the sea, shown in Pritchett's picture, and, landing on a platform, pass to the temple. The island is not allowed to be cultivated, or any one to be buried in it. It abounds in a sort of small red deer, which are perfectly tame, and wander about the streets, going in and out of the houses, for they are considered

31st Dec.

sacred, and no one dares to touch or hunt them. We visited the temples in the afternoon, and were shown over them by the priests, having our feet first covered with carpet slippers. They also showed us the contents of the treasure-house, suits of armour, and enormous two-handed swords, presented by the celebrated emperors and warriors of bygone days. Having had tea and sweetmeats, we went out to look about in the funny narrow streets. The special industry here is wood-carving, and the things were certainly most tempting, but it seemed a puzzle how the people could make a living, as all the shops seemed to contain the same goods, and it is only to pilgrims or chance visitors that they can ever sell anything. We were told that the *Wanderer* was the first foreign vessel that had ever entered this sacred place, and judging from the curiosity exhibited by the troops of people who followed us about this seemed probable.

1882. 1st Jan. Yesterday we had sent up to Hirosima to order jinrickshas, and at 1 A.M. landed on the mainland, and found the little carriages, fourteen in number, awaiting us, each drawn by two men. The town was twelve miles off, but the road was very good, the only traffic being the jinrickshas, foot passengers, and pack animals which were ponies, and a little breed of oxen, with straw shoes tied round their ankles. The men who drew us jogged along at a capital pace of not less than six miles an hour; their wind and endurance is really wonderful. On reaching the town we were met by a gentleman sent by the governor, who took us to a very pretty tea house, where we had hot tea, a tablespoonful at a time in a tiny cup, sponge cake and sweets, kindly provided by the governor, who also sent us two large baskets of oysters, and two boxes of dried fruits. The natives

swarmed about us, and if we stopped to look at anything we were at once in the midst of a crowd, who pointed and talked at us, as evidently great curiosities. Mrs. Lambert's sable cloak with little tails excited a great deal of wonder; the men and boys lifted up the little tails, stroked the fur, turned it up to look inside, and were evidently much exercised as to what sort of an animal it was made from. Starting for the long drive or pull back the men seemed as fresh as ever, keeping up a trot nearly the whole way, with a chanting sort of song. In passing through the small towns, the drawers of Mrs. Lambert's jinricksha, leading the van, would put on an extra spurt, which was taken up by the whole line, and we certainly had the satisfaction of creating a sensation, doors were thrown open, panels were slipped back, and the inhabitants on all sides rushed to see our procession, as we trundled through the streets. The charge for each jinricksha for this day's work, and their twenty-four-mile run was a dollar, and this was a little more than they asked. The governor had given us an order to procure a couple of the sacred little deer, but we had a good deal of difficulty in getting them, as the people for some time stoutly resisted their being taken away. We left Holy Island at 7 A.M., 2nd Jan.
and crossed Misima Nada again, and at 1 A.M. the pilot informed us that we were coming to a narrow passage between the two islands of Isuwadsi and Nima, through which no ship of our size had ever passed, but that he felt confident he could take us through. Everybody clustered on deck, and until we were quite close no opening could be seen, then we crept through a narrow but deep passage, some eighty yards wide, with a town on each side, the chattering voices of the inhabitants heard on each shore.

Here we anchored and had lunch, the natives coming off to see the ship, and bringing fish for sale. At 2 P.M. we were under way again, the weather bright and beautiful, but very cold. The scenery a succession of islands of all shapes and sizes, little towns and woody hills—in fact, very like steaming about on a Japanese tea-tray. Crossing the Suno Nada we anchored at 11 P.M. in Mozi Harbour in the Strait of Simonosaki, the western outlet of the Inland Sea through which we had now completed our voyage.

## CHAPTER XX.

*HONG KONG.—CANTON AND SINGAPORE.*

At 2.15 P.M. we left Mozi Harbour, and, crossing Korea Straits, passed the island of Tsu Sima, and anchored in Fusau Harbour, Korea, at 5.15 next morning. The harbour is situated between Deer Island and the mainland of Korea, on which there is a small and semi-fortified Japanese settlement. The Corean town is a little way off, the two people do not live together, at which one cannot be surprised, for all we had seen of the Japanese had given us the idea of a very clean and orderly race, whilst what we saw of the Coreans impressed us with the exactly opposite opinion. The men were of finer physique than the Japanese, but the expression of their faces is decidedly evil. They wear loose baggy trousers and jacket, and a most extraordinary hat, the use of which it is difficult to understand, as it certainly will keep off neither sun nor rain, being made of horse hair, in the shape of a Welchwoman's hat, and not fitting the head, but being tied down. We didn't see any of the women, as they never leave the town, into which no strangers are allowed to enter. In the morning some of our party landed with their guns on Deer Island, as there were reports of wild boar, deer, pheasants, and other game, whilst the rest of us visited the town, and called

3rd Jan.

4th Jan.

on the Japanese Consul, who showed us about the place. The ladies of our party were the first English women who had ever landed, so were soon surrounded and followed by a mob, but we had two or three Japanese policemen to follow and keep order. The Japanese Consul and the Chief of the Police came off to lunch with us. About half-past four we

A COREAN.

saw the shooting party coming over a hill on Deer Island, and sent off a boat for them. They had bagged several pheasants, and managed to get a shot or two at deer, but not being up to the ground, and, being unable to make the Coreans they had met with understand a word they said, had wasted most of their day before they came on to good game country; they had the satisfaction, however, of feeling convinced that game was there in plenty, had they known where to go. We steamed down the western channel between Corea and Tsu Sima, through lovely scenery of a smiling varied nature, wooded hills, with trees that had leaves of scarlet and yellow, past little villages nestling in their shadow, amidst sampans and junks, by rocks with trees that clung to them and seemed to thrive without any nourishing soil, under a brilliant sun and sky, and sea calm as a lake. It was a great pull having a pilot with us who knew every nook and cranny, for he took us in and out of all sorts of channels and narrows and sounds, that we could not have visited without such help, but which he knew were well worth seeing. The night was almost as bright as day, and during the morning of the 5th we entered the Strait of Ikutski, and steamed round the beautiful harbour of Yelukuro, where a Portuguese settlement existed in former days. From hence we turned the island of Hirado, and entered the strait between this and the island of Do, and anchored for the night, and all agreed that we had seen nothing in Japan to equal the beauties of this day's scenery. 5th Jan.

We left Hirado at 7 A.M. and steamed along the Hirado strait, passing the islands of Kuro and Oo, and further on Malsima and Kaminosima, casting anchor at 2.15 P.M. in the excellent land-locked harbour of Nagasaki. The entrance to the harbour is beautiful, winding in and out among islands that were clothed with evergreens of every variety, and amongst them shrubs with leaves that seemed bespattered with gold and crimson. The town seemed a brisk and busy place, with churches both Protestant and Roman Catholic. Numbers of ships and steamers were lying in the harbour, the greater number flying the German flag, the one most 6th Jan.

commonly seen on sailing ships in Eastern waters. The next
7th, 8th Jan. two days we got a complete change in the weather, pouring rain and hail, so that we were not able to see anything of
9th Jan. the place, and left on the 9th without any regret. The Japanese gentleman who had been with us ever since we left Kobe was a very quiet, thoughtful man, and used to look on at the children skylarking in wonder. Mrs. Lambert asked him if his sisters ever indulged in pranks and romps, but he said "no, that they do nothing but talk, play no games, and have no music."

Now the working class strike me as a singularly light-hearted, merry people, full of fun, and up to any sort of nonsense; possibly the upper class look upon all this as "bad style," if so, I am afraid the high spirits of our young folk must have condemned them in the eye of Japanese etiquette.
10th Jan. We left Nagasaki at 8 A.M., and found a tumbling sea outside,
11th Jan. and the next day the wind increased to a fresh gale by noon, the ship rolling a good deal; the gale blew all night, and a heavyish sea struck the ship about midships, damaging the
12th Jan. bulwarks. About noon wind and weather began to moderate,
13th Jan. falling during the night almost to calm. At 4 A.M. we sighted the light on Turnabout Island on Lamock, which forms the
14th Jan. entrance to Formosa channel. On the 14th we came in during the night and afternoon for the N.E. monsoon and a heavy following sea, which we carried with us all day. On the afternoon of the 14th we passed Pedro Blanco Island, and saw
15th Jan. many junks, and slowed down to wait for daylight after sighting the light on Cape Aquilar at 9.15 P.M. We discovered that we had carried away one of our chain plates, and that some damage had been done to the bowsprit gear. At 6.20 A.M. we proceeded slowly through the Lymoon

channel, reaching Hong-Kong roads and anchoring at noon. On the roughest day of the journey down we had a second small fire on board, one of the large braziers that had been placed in the poop cabin was upset; no one happened to go in for some time afterwards, for when I went in the cabin was full of smoke, and small flames coming from the floor, which was burnt some inches deep. At Hong Kong we found a number of English men-of-war, amongst them two of our old friends of the detached squadron, several of whom dined with us in the evening.

In the morning Bob and I went ashore and called on various people for whom we had letters, and secured rooms at the Hong-Kong hotel, as it was necessary to send the *Wanderer* into dock for various small repairs she needed. 16th Jan.

We took possession of our rooms to-day at the hotel, and found them lofty, large, and very clean, and the cooking capital; in fact, the accommodation generally was a most pleasant surprise. In the evening we went to a capital ball given by the Tarantula Club in a magnificent ball-room at the city hall. 17th Jan.

Mrs. Lambert, with the young folk, went off to the shops full of carved sandal wood and ivory goods, and found the boys, from their Japanese experience, had become as keen as any "old clo" man at a deal. In one shop where Willy distinguished himself by the smallness of his offers, the owner said to Mrs. Lambert, "Sharp boy that, too much savey—nice boy." We lunched with Mrs. Johnson at her house, which stands in a beautiful park outside the town, and afterwards played lawn tennis on one of the prettiest courts, surrounded with trees, that I ever saw. The view from the house over the bay is magnificent. 18th Jan.

19th and 20th Jan. During these two days the ladies employed themselves in visiting the shops and letter-writing, whilst Bob and I went over to Kowloon in a steam launch to see how the repairs to the ship were going on. We found about a hundred workmen employed scraping her sides, previous to caulking and painting; everything looked in a terrible mess. On being docked we found it necessary to strip the copper off her stern port, as it had been worn thin and into holes, in some places by the wash from the screw. That was all that was necessary to be done below the copper line. The decks and upperworks had been dried up by the long spell of hot weather we had had, the paint in blisters, and the standing rigging wanted overhauling and tightening, after two or three of the chain-plates which had been carried away were replaced, and lastly the running-gear and sails which had been most in use were found to be badly chafed in places, so there was plenty to do.

21st Jan. We left Hong-Kong at 8 A.M. in the steamer, *Ichang*, for Canton, taking besides our own party three officers of the *Tourmaline*, one being a very old friend. The steamer was large, clean, and comfortable, all the cabins being on the upper deck. Going up the Pearl river we passed through perfect fleets of junks and sampans, otherwise the river scenery was dull and uninteresting, with flat banks. We reached Canton at 2 P.M., and were struck by what is, I believe, the common feature of all Chinese waterside towns, namely, the enormous population, who are born, live, and die in the sampans; they mass along the shore, so that you can hardly distinguish houses from boats. We anchored in the thick of them, and the whole place seemed to swarm with life, it reminded one very much of stirring up an ant-hill.

The women work as hard as the men; you see the mother, with her baby lashed to her back, pulling away at one of the oars, or punting in shallow water, and helped by any of the children big enough to lend a hand.

Soon after our arrival we took a turn through the town, carried in chairs on the shoulders of two Chinamen, one in front and one behind. The streets were narrow and very filthy. We found the ivory and silk shops very inferior to what we had seen in Japan. In fact, there can be no doubt that to see China with an unprejudiced eye, it should be visited before and not after Japan, where everything is scrupulously clean. Yet we were told that Canton and its inhabitants are considered to be more cleanly and decent than any other town in China, a fact we found some difficulty in believing. We came back to the *Ichang* at sunset, not being tempted to stay at any of the hotels, and made arrangements with the owners and captain to sleep and get our meals on board until Monday morning, when she was to return to Hong-Kong. We were very comfortable on board, **22nd Jan.** and the captain most politely gave up his own cabin and went ashore to sleep, so, with four other small cabins and a bed or two in the saloon, we managed very well. At night the river and banks presented a busy scene, lights twinkled from hundreds of boats and from every shop in the town, every one of which has a little niche in which is an idol with a burning joss-stick in front. The people in the boats kept up a perpetual jabber and chatter, with occasional shouts, enlivened now and then by the beating of tom-toms; the row never seemed to cease day or night. During the day we visited several temples, all far inferior to those of Japan, and all in a state of utter dirt. The most curious is the

temple of the five hundred genii, in which there are five hundred colossal images. The five-storied pagoda is also interesting from its historical associations. It stands on a hill close to the old walls of Canton, and was the spot whence Mandarin and Tartar generals issued their orders, and directed the operations for the defence of the city against the combined armies of England and France. A good view is obtained of the town and country from this place. We had lunch, which we brought with us, in a temple where a large bell is shown pierced with a hole, said to have been made by a shot during the attack. We came back to the steamer, much pleased to be out of the filthy sights and smells,
23rd Jan. and no more in love with Canton than before. Next morning, at 7 A.M., we went on board the steamer *Kiang-Ping*, and went up to Macao, where we arrived at 2 P.M. It is a neat little town, with barracks, hospitals, and churches, but, although it is said to be the oldest European settlement in Chinese waters, it is evidently in a state of mild decay. We visited the tomb of Camoens, and other objects of interest, and coming back to the hotel dined at the *table d'hôte*,
24th Jan. meeting several Portuguese officers. On the following day we left at 8 A.M. in the steamer *White Cloud* for Hong-Kong, and arrived at twelve o'clock and remained at the hotel until
2nd Feb. the 2nd, when, our ship having completed her refit, we were able to get back again to the great delight of all, to our old quarters. The time spent in Hong-Kong, however, was pleasant enough; there was plenty to see, and plenty of amusements, amongst them, an amateur performance of the *Pirates of Penzance*, a capital concert at the German club, and the French play. Mr. Ah-Yon a Chinese gentleman, entertained some of us at dinner on the 30th of January.

The bill of fare, being somewhat of a curiosity, I brought away, and here it is :—

DINNER BILL OF FARE.

**Held at HANG FAH LAU, No. 281, Queen's Road,**
***on 30th January 1882.***

**1.—Turtle Soup.**
**2.—Rock fish.**
**3.—Bird's nest with Minced fowl & Ham.**
**4.—Pegions' eggs.**
**5.—Fish fin with Minced fowl.**
**6.—Stewed sliced pegions.**
**7.—Stewed prawns with bamboo shoots.**
**8.—Stewed sliced fowl.**
**9.—Sweet Lotus seed broth.**
**10.—Various pastry.**
**11.—Fragrant fungus.**
**12.—Minced Quails & Ham.**
**13.—Goose's throats.**
**Fruits, Coffee, various sweetmeats.**

AH-YON.

The games at shuttlecock in the streets are very amusing to watch, for any one who is passing by seems entitled to join the game. When once the shuttlecock is thrown up into the air it is kept up with the feet of the players—not with a battledore—and they strike it with the broad sole, sometimes like "Old Joe" kicking up before, and sometimes kicking up behind. You see a mother or nurse, with a baby strapped to her back, suddenly leave the side of the road, and dart into the middle to join the players, and, having satisfied herself with a few kicks, go on her way again. These were three busy days, **3rd, 4th, 5th Feb.**
paying and receiving visits, amongst other visitors Prince Louis of Battenberg came one morning to breakfast. On the **6th Feb.**
6th we had a sailing match. I offered a cup if the men-of-war would send representatives, and we had twenty-one entries.

The start was at 2 P.M. from a line laid between our ship and the Portuguese transport lying near us, the course having been arranged by Mr. Cecil Burney, the umpire, Captain Jones of the *Victor Emmanuel*, and the starter, Mr. Montgomerie, of the *Carysfort*. It was a very pretty start, although the wind was light, but, on turning the given point to beat back, a nice breeze sprang up, and the good boats soon

SHUTTLECOCK, HONG KONG.

began to show their quality. The French admiral's barge from the *Themis* frigate started first favourite, took the lead from the first, and kept it to the finish. She was a fine boat with beautiful sails, and was very well handled. One of the *Inconstant* boats was second, and Bob and I came in a good fourth in the galley. When all were in, Mrs. Lambert presented the cup to the Frenchman, who, I think, was much pleased, as he came next day to consult us as to the inscription. We began 7th Feb. early in the morning to leave our P.P.C. cards and say goodbye to our many kind friends ashore, and, on coming back to the ship, found Mr. Ah-Yon waiting on board to say goodbye. He had brought Mrs. Lambert a parting present in the shape of ornaments similar to those that are placed before the god on their home altars; these are difficult to describe, being groups of little figures of men, with flowers, feathers, and glass, very bright and glittering, and two boxes of dried lychees. We started at 11.30 A.M., and were followed for some little distance by Mr. Ah-Yon in his boat. Some of his crew held out a long bamboo, with a sort of basket at the end in which a fire was burning; into this they kept throwing handfuls of crackers,—no doubt it was meant as a great compliment. Very light winds and calms in the 8th Feb. morning, but towards noon the wind freshened from the N.E., and we feathered the screw and went on under sail; the three next days we got beautiful weather with a smooth sea 9th, 10th, 11th Feb. and light winds, making from six to seven knots an hour. Towards evening on the 12th the wind got stronger, but we 12th Feb. managed to hold our Sunday service, although the decks were rather wet, and we were doing between nine and ten knots. Next day we got a fairly good breeze, but on the 14th the 13th, 14th Feb. wind fell away, and we consequently felt the heat a good deal. Wind still failing and the heat increasing, at noon we 15th Feb. had to begin steaming again, and anchored in Singapore roads 16th Feb. at 5.30 P.M. to-day. Early on the following morning we went 17th Feb. ashore to see the fish and fruit market, and found abundance of bananas, cocoa-nuts, mangosteens, and oranges. There was nothing very remarkable in the way of fish, except the tubs full of *bêche de mer* for the Chinese, looking like enormous slugs. The town has no striking points about it. At noon we went to call on the governor, Sir Frederick Weld. The government house stands on rising ground a little way out of the town, and is surrounded by beautifully kept grounds, looking like an English park, except for the presence of the fan palms, banyan, and bright flowering trees. We drove under a fine portico, and were ushered into a large hall by several servants dressed in white with scarlet and gold waist-bands, and belts across the chest—very handsome dark fellows. Whilst waiting in a large and lofty room, punkahs waved over our heads, a new sensation to us. We were most kindly received and stayed to lunch, and later on, when it was cooler, the governor took us in his four-in-hand for a very pleasant drive over excellent roads to the reservoir which supplies the town, and to the botanical gardens, then down to the town and landing-place, from whence we hurried on board, being rather late for dinner, to which we expected Captain Sandilands of H.M.S. *Mosquito*.

18th Feb. Sir Frederick, Lady Weld, three of their daughters, and Captain Tannard, their aide-de-camp, came to lunch, and spent some time going over the ship. The governor, who is a nephew of the late Mr. Weld, who owned the celebrated *Alarm* cutter, had seen a good deal of yachting, and took more interest than many of our visitors in looking over the different parts of the ship. We found that we had seen Lady Weld's brother, Mr. Rudolph de Lisle, in Chili, as he was on board the *Shannon*, and that we had mutual friends in Staffordshire.

19th Feb. We had service as usual. The congregation was swelled by the presence of Lieutenant Story, and twelve men from

20th Feb. the *Mosquito*, the day hot and sultry. The sportsmen of our

party, undaunted by their previous failures, went off in a very comfortable steam launch to Alligator Island, but got no sport, and only saw traces of wild pig. In the evening we went to Lady Weld's "at home," and spent a very pleasant evening. Shortly after our arrival at Singapore we had received an invitation from the Maharajah of Johore to go and spend a day with him. So on the 21st we left Singapore 21st Feb.
at 10 A.M., and, steering through the beautiful straits, arrived and anchored opposite the palace about 3 P.M. Shortly after the secretary came off, and we went ashore with him to pay our respects to his highness, who is a very courteous and pleasant gentleman. We entered the palace through a very fine hall paved with marble, and up a broad white marble staircase into a very large dining-room, full of numbers of beautiful things. There were many crystal candelabræ ten feet high; from this room we passed through a splendid ball-room to a wide, marble-paved corridor, overlooking a beautiful garden. We were afraid to stay very long, for thunder was muttering in the distance, and we were anxious to get back to the yacht before a storm broke. In the evening we returned, a party of eight, to dine with the hospitable Maharajah, and found that he had some guests staying with him, Lord and Lady Harris, Lord Zouche, Mr. Tufnell, and Mr. and Mrs. Currie. The dinner was most excellent, the punkahs waving over us kept a most comfortable temperature, and the Malay attendants dressed in white with gold and crimson caps and twisted waist-belts added not a little to the picturesque effect of the whole scene. During dinner an excellent band composed of Malay musicians, but led by a German bandmaster, played in the ante-room. After dinner the ladies went into the drawing-room until the gentlemen came and invited them to the

billiard-room. The Maharajah has a house also in Singapore, living sometimes there and sometimes at Johore. Unfortunately we did not see the Maharanee, as she was at the former place. His highness was most kind, and offered us beds, but we preferred going back to our own quarters. At 22nd Feb. 7 A.M. next morning the gentlemen of our party and the two boys went ashore and drove to see the Maharajah's new plantations of coffee, tea, gambier, and pepper, which looked thriving, and promised good returns in the course of a few years. The Maharajah's object in laying out these plantations is to try and induce his subjects to follow his example and make the state more productive. The party were back to breakfast with the news that the Maharajah and all his guests, who were going back to Singapore, would join our party on the yacht, and at ten o'clock they came off. In a second boat came a present for us of a cheetah in a strong, iron-bound cage—a most ferocious-looking animal. All being on board, we left at 10.30 and had a very pleasant day, reaching Singapore at 4 P.M., when the party quickly broke up. Lord and Lady Harris were off by a steamer at 6 P.M. for Siam, and Lord Zouche to Australia. As we neared Singapore we sighted the detached squadron (the *Inconstant, Tourmaline, Carysfort*) beating up for Singapore. The *Inconstant*, being much the fastest ship under canvas, was leading by some miles. We steamed round her to bid good-bye to our old friends, whom we could hardly hope to meet again on this voyage, unless we fell in with them at Gibraltar, which was not very likely, and, having landed our guests, at 5.30 P.M. we said good-bye to Singapore and proceeded on our homeward course.

## CHAPTER XXI.

*FROM SINGAPORE TO CEYLON, SUEZ, CAIRO, JERUSALEM, AND BEYROUT.*

By noon to-day we had run, under easy steam and sail, 23rd Feb. 125 miles, and five miles more brought us to Malacca. We landed after lunch to see this quaint old place. The acting lieutenant-governor, to whom Sir Frederick Weld had given us an introduction, was very kind and obliging, and, procuring some carriages for us, sent an attendant to show us what was most worth seeing in the little place, amongst these being a 24th Feb. picturesque Dutch gate forming one of the entrances to the fortifications. At 6 P.M. we were on board and off again on 25th Feb. our voyage. Next day a nice breeze sprung up from the 26th Feb. north-east, and about ten o'clock we stopped steaming, and went on under sail. By 7 A.M. we had cleared the straits 27th Feb. and shaped our course for Point de Galle. The wind was light with occasional squalls, and next day was rather baffling, but the weather fine and pleasant. From the 28th to March 3rd we got fine weather, with light breezes and smooth water, the ladies thoroughly enjoying themselves. At 4 A.M. we 4th March. began to steam, and at 2 P.M. were lying in Galle Roads, where we didn't stay long, for, having just landed and found no letters, we immediately went on to Colombo, where we anchored at 8 A.M. next morning, after a perfectly calm

5th March. passage. After service (it being Sunday) we got our batch of letters and papers. There is a backwater here, which makes the harbour very secure, with excellent anchorage. There seemed to be a busy trade going on, for coasting brigs passed in and out, laden with country produce, and manned by Cingalese. Large steamers also, under the French, German, Dutch, Italian, and English flags, were daily arriving and leaving.

6th March. A native juggler came on board and amused us very much with his clever tricks; it was a wet afternoon, so we were

MOONFISH, CEYLON.

7th March. glad of some amusement. The next day we were up soon after five o'clock, in bright moonlight, and after getting some breakfast, went ashore to start by the seven o'clock train for Kandy. The first part of the journey was through flat country, marshy and wet—with rice fields, and luxuriant vegetation. Then we began gradually to climb up the sides of hills, pop in and out of tunnels, and get higher and higher, until we got to a part called "Sensation Point"; from here we could look out of the carriage windows, down a yawning precipice, and away to the valley far below us, dotted about with the planters' houses, and plantations of coffee and cinchona, interspersed with roads and rivers. It was a pleasant cloudy day, and when we reached Kandy at eleven, we found it quite cool and breezy. The hotel stands in a beautiful position, overlooking a lovely lake surrounded by hills, thickly grown with trees in front. In a green inclosure stands an old temple of massive construction, and of a beautiful grey colour. After lunch we went off to the botanic gardens, a pleasant drive some little way out of the town, passing some houses where a native wedding was going on. A large arch was in front of the door made of green bamboo, decorated with bunches of fruit. A river flows through the middle of the gardens, which are most beautiful, not so much for their flowers, as for the beauty of the trees and coloured leaves, palms of every variety, and yellow and green stemmed bamboos. Close by is an old Buddhist temple, where offerings of flowers lay before the shrine, and outside, in the verandah, hung pictures representing the punishment of the wicked, some being burnt, others being cut to pieces or eaten by wild beasts. There is also a very fine collection of Cingalese books, beautifully written on strips of palm leaves, bound in bands of painted and engraved wood, or silver, which were shown us by an old priest dressed in a whisp of yellow cotton. There was a *table d'hôte* at the hotel, at which we dined, and found our rooms very comfortable, and left next day for Colombo by the 2 P.M. train, and got on board again in time for dinner, to find the coaling all done. We had very few **8th March.** visitors, as nearly all the residents had gone up into the hills, to get out of the hot weather. After making some purchases **9th March.** next morning as we came back to the landing-place, we found a crowd collected round a number of men who, dressed in scarlet and carrying banners, were waiting with a band of music to receive an embassy from the Maldive islands, which had arrived to pay the annual tribute. So we waited with the other lookers on. Presently guns began to fire, and the ambassador landed from a small vessel, with a retinue, who carried before him rolls of mats, jars of honey, and bags of cowries; over his head was carried a piece of folded muslin containing probably some coins.

10th March. We started at 4 P.M., and at noon on the 10th had done 150 miles, and a nice fresh breeze from the N.E. springing

11th March. up, we stopped steaming and got under sail. To-day the wind was very light and baffling, and, coming round right ahead suddenly at 10 A.M., the square and studding sails had to be taken in; whilst this was being done, John Gates, O.S. fell from the futtock shrouds, and, making a series of backward somersaults, bounded off the starboard lower rigging, and fell overboard clear of the ship. Mr. Tyacke, the chief officer, threw a life-buoy close to him, and at the same time the vessel was thrown aback, and the life-cutter manned and lowered. Powell, the coxswain, was in his bunk after his watch, but was in the boat without waiting to dress, almost at the first sound of "man overboard." In fact, the whole thing was done with most praiseworthy smartness, as it did not take more than ten minutes from the time the order was given to " lower away," to the time the cutter was hoisted up again, and Gates safe and sound on deck. He was a good deal bruised, but had never lost his senses, and was able to swim. *Next day* a large shark was seen close to us—thank God he had come a day too late.

12th to 20th March. At 6 A.M. we sighted Minnikoi, one of the Maldives, a very pretty atoll—the eastern side is overgrown with cocoa-nut palms, and here we saw some huts, but the western rim is

just a low coral bank without vegetation. From to-day until
the 20th we had an almost flat calm, with very light airs
and hot weather, but about midnight on the 20th it became
dark and overcast, with a heavy shower of rain, and it was
so thick that we could not see the land until 11 A.M., and at 21st March.
noon we were eleven miles off Aden. We kept on our course
for Perim, as there was a chance, if we touched at Aden, of
being put into quarantine, and at 4.50 P.M. were abeam of 22nd March.
Jebel Kan, sighting Perim light at 9.20, and passed through
Bab-el-Mandeb at 11 P.M. I feel that the voyage through
the Red Sea is known to such a vast number of English
ladies and gentlemen, that I will not detain my readers with
an account of our passage through it, but go on at once to
Suez, where we arrived on the 28th about noon. 28th March.

Next morning we were ready for a start by ten o'clock, but 29th March.
three steamers being on their way through the canal we had
to wait for them. The first one was the *Crocodile*, full of
troops for India. About twelve, however, we did make a start,
and got up the canal about four miles when we had to wait
for two hours to let a fleet of seven steamers pass us. At
4 P.M. we got on again, and as far as the Little Bitter Lake,
fifteen miles from Suez, where we arrived at 6 P.M. and
moored for the night, in company with an English and
German steamer. Next morning we started at 5.30 A.M., 30th March.
and steamed full speed through the Great Bitter Lake in
hopes of getting to Ismailia without any further stoppage,
and catch the eleven o'clock train for Cairo; but at the end of
the lake we were met with signals to stop, and wait again for
more steamers to pass that were coming through the canal;
there was nothing to be done but obey, and make up our
minds to lose a day. We got to Ismailia in the afternoon,

and some of our party went ashore, and came back with oranges, and orange blossoms, but were not much struck with the town. Anchored here we found the yacht *Marquesa*, with the owner, Mr. Kittlewell and his bride on board. She is a nice-looking vessel of 400 tons, but she left almost as soon as we arrived, going on her way to Suez.

31st March. We left next morning by the eleven o'clock train, which we found very full, and had to split up in twos

WATERWHEEL, CAIRO.

and threes, until we reached Zagazig, where we changed and were rather better off. For the first two hours the journey is through barren desert, but after this you get into cultivated lands, irrigated and with fine crops of wheat. We reached Cairo at half-past five, and drove off to the Hotel du Nil, which is very comfortable and quiet, being some little way off the noisy streets and traffic. It has also a pretty garden at the back, where we walked after dinner, which, by the way, was excellent, well cooked, and everything good. We started off at 6 A.M. to-day, a party of sixteen, for the pyramids, over a very fair road shaded by trees on each side, through a most fertile country, teeming with splendid crops of wheat, barley, and grass. The latter was being cut and carried into town by scores of camels and donkeys. I think that even with the photographs, and quantities of books upon the subjects, we had pictured the pyramids as standing on a flat plain, and were surprised to find how hilly the country really is. Close by stands a deserted house, which was built for the accommodation of the Prince and Princess of Wales, and looks strangely out of place. We got back again to Cairo at 11.30 A.M. *in a shower of rain.* "Is there any need to take an umbrella?" said Mrs. Lambert before starting, to one of our party, who, from having been in Egypt before was accredited with weather-wisdom, and the prophet had answered, "It rains about once in five years." It was very curious that we should have been here during the fifth year. After lunch at the hotel we started off again to do some sight-seeing, through old Cairo and to the citadel and Mohamed Ali's mosque. The rain poured, and we were glad to hurry into our carriages and get back to the hotel. Next day we devoted the morning to the museum, Omar's mosque, and the old Kopt church, and the afternoon to a drive to the Shubra avenue—the Rotten Row of Cairo—where we saw the Khedive, escorted by a squadron of cavalry, and many nice carriages and smart-looking people. We were fortunate in securing the services of Louis Mansour as our dragoman, for a more obliging, active, and useful guide it is impossible to imagine; he arranged everything for us in the most comfortable manner, managing the hire of horses and vehicles, providing food,

1st April.

2nd April.

keeping the natives pleasant, and preserving us from extortion. 3rd April. We left for Ismailia again to-day, taking Mansour with us, and stopped at Zagazig for lunch, and meeting a long train of pilgrims returning from Mecca, most of whom were Algerians. We got to Ismailia at half-past five, and were glad to get on board again. 4th April. Next morning we went on by the canal again to Port Said, preceded by an Egyptian steamer full of pilgrims, which kept continually grounding on one side or other of the canal, and delaying our passage in a most irritating fashion. At last, however, she stuck near a *gare*, and by applying to the authorities we got leave to pass her, and so saved a day and got to Port Said at 6 P.M. It was a most uncomfortable day, with a strong wind blowing right across the canal, and occasional showers of rain. Our temper was not improved on getting to Port Said, to find that the Arabs had all struck work, and that we must *coal* with our own crew. 5th, 6th April. The next two days were spent in true discomfort and dirt, for all around us ships were either discharging or taking in coals, and clouds of dust and grit pervaded the air. We had not been in such a mess since we left home; at 5 P.M. on the 6th we had finished, and got out to sea, declaring that Port Said was the filthiest place we had visited. 7th April. We got to Jaffa at 8 A.M. on the following morning, weather rough and blowing hard, and found the *Bacchante* there. One of her officers who came on board told us that a party from their ship who had gone to Jerusalem, on getting back to Jaffa had found her gone, as she was obliged to put to sea, the harbour being so bad, and the weather so nasty. Another party, however, were going to start at once: the young princes also had gone for a tour inland, and were to join the ship again at Beyrout.

There was a heavy sea running into the harbour, and it was with some difficulty that we got the ladies into the galley. About half-way towards shore we were met by a large Arab boat provided by Mansour, as it would have been dangerous to have landed from our own boat, and into this we bundled and tumbled and scrambled. About a dozen men rowed, standing up and singing a chanting song, that reminded us of the negroes at Gaboon. On landing we walked through narrow streets crowded with Asiatics from all parts, and full of camels and donkeys, to Howard's hotel, where we lunched. About one we made our start for Jerusalem. Mrs. Lambert, Miss Power, and I, travelled in a bathing machine, or vehicle very much resembling one, drawn by three horses abreast, whilst the two girls, Bob and the boys, with the rest of our party, rode on horseback, with White and Mansour to escort them. The luggage was carried on pack-horses. Outside the town, the groves of orange and lemon trees, full of blossom and fruit, were most fragrant and delicious—they could be smelt from the yacht when the wind was off shore. In the distance, as we drove along, we saw the town of Lydda, "which is nigh unto Joppa." About five in the evening we stopped at a large stone house, another of Howard's hotels, in the valley of Ajalon. Here we gladly dismounted—we from the bathing machine with a sore and bruised feeling, having been well jolted over the eighteen miles on the hard seats, and those on horseback, not having ridden for some time, felt as if they had had quite enough. After a good night's rest, for the beds were comfortable, we started next morning 8th April. at 8 A.M. on a very much worse road than previously, and got into the hill country, with wild flowers in abundance, mignonette, cyclamen, and gorse, a few fig trees and olives in the valleys. After about twelve miles, on coming down a steep zigzag road, we found our lunch under a grove of olives prettily laid out by Mansour, who had sent it on before. Rugs were laid out on the ground—mugs with wild flowers, oranges, dates, raisins, figs, were scattered about, with more solid features in the shape of fowls and cold meat. It was a most enjoyable picnic, after which we went on again climbing and winding in among the hills until about 3 P.M., when Jerusalem was in sight. We entered by the Damascus Gate, and went to the Damascus hotel, built of stone, with a flat roof, and steps open to the air, that lead from one part of the building to another. Just after we arrived there was a great shouting and tumult in the street, crowds of people rushing from the Holy Sepulchre church. They had just obtained the holy fire, as it is called. The struggle amongst the various sects of Greeks, Armenians, and Copts, as to who shall first get their candles lighted, is frantic; we heard that one man had been killed. We went out at once to see this church, and found it crammed, so stood behind a line of soldiers, and watched a procession march round the sepulchre, which is in the centre of the church, each sect headed by its bishop splendidly robed in cloth of gold, with jewelled mitres, and priests swinging censers and chanting. On the next day, which was Easter Sunday, we again visited the church of the Holy Sepulchre, furnished with a janisary from the consulate, who had a curved sword at his side and carried a whip in his hand. The decorations were gorgeous, and we stood for some time listening to priests in various pulpits, each in turn reading a verse of scripture in many languages, including English, but with a very strong accent. Afterwards we walked outside

9th April.

the city walls, seeing the valley of Jehoshaphat, the tomb of Absalom, the Garden of Gethsemane, and the Mount of Olives. From hence to the Jews' quarter, through lanes of the filthiest description, and then back to the hotel; the next two days we visited the Mosque of Omar and the other places of interest, which I do not attempt to describe or enlarge upon, feeling that it is not in a journal of this description that readers will look for an account of Jerusalem, with its sacred associations. I think that we all felt that it

10th, 11th April.

DÆDALUS LIGHTHOUSE, RED SEA.

is better to read of Jerusalem than to visit it. Jerusalem has been captured some twenty times or more ; on several of these occasions it has been destroyed, and so much rubbish from the ruined buildings has been accumulated, that it is very difficult to make out exactly the situation of the various localities, which must be buried far below the present city. We left Jerusalem at 8 A.M., walking to the Damascus Gate, where we got into our bathing machine, for which we had

12th April.

had some cushions made, increasing our comfort very materially, and arriving at Latrun, spent the night there as before, leaving for Jaffa next day. It turned out a very rough day—torrents of hail and rain thoroughly drenching the riding party. When we got back to Jaffa, we found it was so rough that there seemed no chance of being able to get off to the *Wanderer;* she had got steam up, afraid to remain at anchor, and was rolling and pitching in a way that was suggestive to some of anything but pleasant experiences. However, about 4 P.M. there was a slight lull, of which we took advantage, and all got off in a large Arab boat, and after a good buffeting reached the yacht, into which we were hoisted in a chair which had been got ready. We got under way at once, the ship rolling heavily throughout the night,
14th April. and at 8.30 A.M. next morning anchored at Beyrout, finding very cold weather and the mountains of Lebanon white with snow.

COWFISH, CEYLON.

## CHAPTER XXII.

*DAMASCUS, CYPRUS, CONSTANTINOPLE, MALTA, AND SYRACUSE.*

LOUIS MANSOUR went ashore as soon as we had anchored to make arrangements for the trip to Damascus, and came back again with the news that the diligence would leave at 4.30 A.M. next morning, so after dinner we all went to the hotel to get as much sleep as possible before our early start. At 3.30 A.M. we were sitting down to breakfast, and at the diligence office by 4 A.M. packing ourselves away in the various compartments. The coach was drawn by three horses and three mules driven abreast, and travels a splendid road that was made by the French during their occupation of Syria in 1860. Their government still remain proprietors of the diligence, and of all wheeled carriages on the road. The driver and conductor were both French-speaking Arabs, and very pleasant, obliging fellows. We soon left the level country and began to ascend the Lebanon range. The pass is 5,000 feet above the sea level. You look down upon the shipping and town of Beyrout until the summit is reached, when the glorious valley of Lebanon spreads out before you on the other side. It is a splendid sight, but needs a far more experienced pen than mine to describe the constantly-changing colours on the mountains, the picturesque

14th April.

15th April.

villages, the rocky knolls, vineyards, and mulberry gar that we gazed on. Descending the range we stopped a village of Shtora to breakfast, after which, crossing the pl the road again ascends for some distance, then pas through a stony valley you come to the river Barada, b known to most of us as "Abana," from hence you get am green lanes, orchards, and gardens. At the last stage white horses were put in, with smarter harness, for our e into Damascus, whose towers and minarets were now visi On arriving at the Hotel Demetri we found it very full, separate parties of tourists having nearly taken every The poor landlady, who spoke only Arabic, was attacked the dragomen, each trying to get the best of everything his own party; in the intervals of bargaining she cons herself with a cigarette or pull at a hubble-bubble. T hotel is comfortable, built in a square, with a centre cou containing a tank, a fountain, and orange trees. There no private sitting-room to be had, so we sat in this courty or in the balcony. We were very glad after our long da drive to go off to the beds which had been arranged for

16th April. and, after a good night's rest, started after breakfast to see great mosque, formerly the Temple of Rimmon, from here the Church of St. John Baptist, which is said to be the old Christian church in the world, although it is now used as Mohammedan mosque. From one of the minarets a mo beautiful view of the city and surrounding country is g over the fertile belt of cultivated land adjoining the tow away to the barren mountains beyond, with Mount Herm a mass of snow. Coming down we walked through th bazaars and the old citadel, with its interesting re-miniscences of Saracens and Crusaders. The streets ar much cleaner and more even than those of Jerusalem. After lunch we sallied out again to visit some houses that are kept up in the old style belonging to wealthy Jews. They had beautiful marble courts, with fountains and flowers, the walls of the rooms decorated with carvings, and the windows with stained glass. One of these houses belongs to the Danish consul, and we were received by a young lady with brown hair and blue eyes, we thought she must be a Dane, but found she was a Syrian Jewess, speaking Arabic, and a little English and French, which she had learnt at an English school in Damascus. We were shown into a room from whence we heard some singing, and on entering found fifteen ladies seated on a divan round a raised daïs, which was covered with fine matting, and on which stood the hookahs they were smoking, with their long tubes covered with bright silks. We all sat down and sweetmeats were handed round. It appeared that this was a visit of ceremony from relations to the consul's sister-in-law who had been married two or three days before. After some singing we came away. The Jews are generally very handsome men here, with fine feature s dark eyes, and beautiful teeth. We took a walk down the street that is called "Straight" to see some silk shops, meeting a funeral with mourners singing at the full-pitch of their voices. The next day was spent in visiting the bazaars 17th April. and making various purchases. The town is very interesting, and the people and costumes present a constant study of what is picturesque and quaint. Indeed, Damascus is a place where one might spend some time with great pleasure.

We left the town at 4.30 A.M. by the diligence, charmed 18th April. with the scenery on our return as we had been going to

Damascus. When within about nine miles of Beyrout one of the hind wheels of the diligence came off, after wobbling about for a little while, we subsided gently on one side, and no one any the worse. A man was sent back to the last post-house and we sauntered on for about an hour and a half, when the diligence overtook us again, and picking us up, we got down to the port and on board again at 8 P.M., pleased to
19th April. find a calm sea. Before leaving Beyrout we went to see the

DAMASCUS DILIGENCE.

wife, and daughter, and mother, of our excellent dragoman, Louis Mansour. He has a very nice house, and the family received us most warmly; being Christians the women of the family do not cover their faces. The old mother was evidently very proud of this good fellow, and could not wish us a better wish than that our sons should be as good as hers. The daughter, a pretty girl of sixteen, gave each of the ladies a coloured handkerchief, such as the women wear on

their heads. We sailed for Famagousta at 5 o'clock in the evening, and had a quiet, pleasant voyage through the night, anchoring next morning at 8 A.M. in Famagousta roads. 20th April.
Captain Gordon, the assistant commissioner, very kindly showed us about this curious city of ruins, for really there does not seem to be an entire house in the place, the inhabitants live in various parts of the massive ruins of churches and noble old buildings that were built by the

LARNACA, CYPRUS.

Venetians, and destroyed by the Turks when they took the island in 1571. One very fine old church has been patched up and is used by the Turks as a mosque, another for stables. We visited the house once inhabited by Cristoforo Moro, a Venetian Governor of Cyprus, said to be the original of Othello. Supplies here were plentiful, eggs 2*s.* 6*d.* a hundred, the best fowls 1*s.* 3*d.* each, and beef 4½*d.* a pound. We left for Larnaca next morning at 5 A.M., and arrived at 10 A.M. 21st April.
after a beautiful run along shore of thirty-five miles, and left

again at 4 P.M. as there seemed nothing worth stopping to see, still steaming along shore in view of a fertile country covered with corn, vines, and mulberry plantations, and studded with many small towns and villages. The next two days we had lovely weather, with sea smooth as glass. On the 23rd we sighted the high land of Asia Minor on our starboard side, and at daybreak had the island of Rhodes on our port bow. We had our Sunday service at 10.30 and anchored off Rhodes between three and four in the afternoon. We passed a quiet afternoon, and did no sight-seeing until the next day, when all went ashore. Marine fortifications surround the old town, which is full of interest to the antiquary. In the olden days the harbours must have been excellent ones, and the ruins of the ponderous piers still exist. The most interesting street is that of the Knights, for over the porches of many of the old houses are still to be seen the escutcheons of different grand masters of Saint John, and the arms of the town appear over the various gateways. High above any other point in the town, at the top of the "Street of the Knights," stands a mosque which occupies almost the same site as that on which used to stand the Knights' Church. This was blown up some years ago by a flash of lightning igniting a considerable amount of gunpowder stored in the vaults. It must have been there for generations, the Turkish authorities not being aware of its existence. Sad relics of the beauty of the church lie strewn around, marble slabs and shapely columns. We climbed a tower standing on the eminence, and, favoured by the brilliant day and bright, clear atmosphere, got a perfect view of the town and its three harbours, the beautifully timbered country and the sea beyond. The Jews in the town, of whom there are many,

22nd and 23rd April.

surprised us by talking Spanish. We left at 4 P.M., and passed during the night between Rhodes and Cape Aleppo, crossing the Gulf of Kos, and through Kos Channel, passing through groups of islands, by Patmos, with its church on the top of a hill, and later on between Samos and Furin, in perfect yachting weather. During the night we had to change our course again to pass up the Straits of Chios, and entering the Gulf of Smyrna anchored at 8 A.M. in its beautiful harbour. It is a large and improving-looking place, with a fine new dock and walled-in harbour. Inside the dock lay a schooner-yacht flying Austro-Hungarian colours and the R.Y.S. burgee at the main. She proved to be the *Menai*, 175 tons, owned by Prince Louis de Bourbon, who married an Austrian princess. He, poor fellow, was laid up with fever and was taken ashore next day. We landed in the course of the morning to make arrangements for a trip to Ephesus. Later on in the day some of our party ascended Mount Pagus, from which a magnificent view of town and harbour was obtained.

25th April.

26th April.

Next day some of our party left by special train at 7.45 A.M. for Ephesus, and found the ruins most interesting. Our guide was a man who was evidently anxious to show us everything of which he or we had ever heard, and in a very short space of time he had pointed out the Church of St. John, a Saracenic mosque, the site of the Temple of Diana, the town clerk's house, the Stadium, Temple of the Sun, the Theatre, scene of the Demetrius riot, the Temples of Cæsar and Hercules, and St. Luke's church and grave. We got back to port and on board again by 4 P.M. The schooner yacht, R.Y.S. *Erminia*, with her owner, Major Murray, on board, came in during the day, and he and Mrs. Murray came to lunch. Having got our party from Ephesus on board,

27th April.

we proceeded on our way to the Dardanelles. Rounding
28th April. Cape Hydra during the night we shaped our course up
Mytelene Channel, and through the Mousselim Channel,
rounded Cape Baba with its pretty town and fort, and
crossing Besika Bay, entered the Dardanelles at 11 A.M.,
and anchored in Sari Siglar Bay, so that we might get the
necessary firman from Constantinople. On landing to call
on our consul we heard to our dismay that there was but
little hope of getting the permission to enter for three or four
days, so came back to the ship in a very crestfallen condition.
29th April. We made another effort next day, and went ashore in the
afternoon to see if our consul could give us any comfort.
As none, however, was to be got from him, I telegraphed
to Lord Dufferin, to which he kindly replied as follows: "I
am doing my best to obtain the permission, and hope to get
it to-morrow or the day after." We went back again to the
ship, threatening to spite ourselves by not going to Constantinople at all, but on direct to Athens. However, we
made up our minds to wait a little longer, and so spent
30th April. a quiet day on board on Sunday.

1st May. To-day the monotony of waiting was broken by a little
incident and by seeing others put in the same position as
ourselves. A steam yacht came briskly up, and passed us
as if going straight in, when boom! boom! boom! Three
guns from the batteries on either side were fired! Boom!
boom! Two more from the fort in front; and the yacht
stopped, and came to anchor near us. She was the *Zingara*,
R.T.Y.C., with Lord and Lady Wolverton and some friends
on board. We went to call upon them, and found they had
telegraphed from Athens for permission, which they had
counted on being granted to enter at once. We had the

questionable satisfaction of depressing their spirits by telling them we had been waiting three days. Late at night we got another telegram from Lord Dufferin saying they still awaited the Sultan's answer. Next day the surroundings 2nd May. were rather more lively, as three Dutch men-of-war came in, anchoring near us, and there was a good deal of saluting and playing of bands. They were followed by the *Erminia*, and what with exchanging visits, and displaying our curiosities, we forgot our weariness. At 5 P.M. Mr. Maling, the consul's son, came off with the long-looked-for authority, and we had just time to get up steam and under way so as to pass the forts before sunset. The *Zingara* came in also, but the *Erminia*, being a sailing yacht, couldn't manage it in time, and we also left the Dutch men-of-war anchored off Chanak awaiting their firman.

We anchored in Constantinople Roads at 9 A.M., and in 3rd May. the afternoon went off to thank Lord Dufferin for his kind efforts. The first impression we got of the town was decidedly disappointing, for it was enveloped in smoke, and *London* rose in one's thoughts. Later in the day it cleared off, and we were able to see the hills, and mosques, the minarets and domes of the city; but there is a great traffic of steamers from one part to another, pouring forth volumes of black smoke from their funnels.

We went off sight-seeing, and visited the Mosque of Saint 4th May. Sophia, it is larger than the Mosque of Omar, but not nearly so beautiful—in fact the mosques of Cairo, Damascus, and Jerusalem, are, we thought, much finer than those of Constantinople. From thence we went to see a sort of Turkish Madame Tussaud's—plaster figures dressed in old Turkish costumes. Then to the great cistern, supported by its thousand and one columns. After lunch we went out again to see the bazaars, and did not find them very tempting after all we had seen elsewhere. On coming on board we found Captain Jolliffe of the *Antelope*, and an old friend of the *Repulse*, Captain Grenfell (now of the *Cockatrice*), who told us the sad story of Captain Selby's murder. As we sat on deck the voices of those who call to prayer came floating over the water in musical tones from the men appearing on the little balconies of the many minarets.

5th May. Next day, escorted by these two gentlemen, we went to see the Sultan go to mosque. It was a Friday, which is a sacred day with Mohammedans. We went in a steam launch, and, passing one of the many palaces, were shown a Turk, who was sitting at a window. He was supposed to be the ex-Sultan, who is a madman; possibly the man we saw was only his representative. Landing near a large barrack we walked through lines of soldiers to a road leading to the palace. A little further on we came to the guard house, where we stood on the steps with numerous other strangers to see the procession. The road was lined with soldiers and attendants with silver censers of incense to perfume the air. Presently a bugle sounded, the band played, and the soldiers presented arms, and three little princes came out of the guard-house to join in saluting the Sultan. These little fellows were all dressed in uniform, the eldest looked about ten, the youngest a charming little boy of five. Two were in naval, and one in military uniform, but all wearing the fez and a diamond star on their coats. Then the great man appeared, preceded and followed by crowds of officers in brilliant uniforms, amongst them we saw Baker Pasha. The Sultan rode a most beautiful grey horse with gold chains on his neck, golden bit, and gorgeous trappings, but he looked ill and miserable, and lives, they say, in constant dread of assassination. After seeing the Sultan safe to the mosque, Hobart Pasha joined our party. Then came a march-past of the troops, at which the Sultan looked from a window in the mosque, peeping through Venetian blinds, so that no one might know at which window he was stationed. It was a very gay sight. There was one black regiment, of very fine, tall men, and last of all came a troop of Circassians in handsome uniforms. When all had passed the Sultan came out of the mosque and got into a beautiful little carriage with a pair of grey horses; he took the reins himself, whilst grooms in blue and gold ran by the side, and so he disappeared through the gates into his park, not to come out again until next Friday. We then got into carriages which our guide had got for us, and went to see the dancing dervishes in a large mosque with the central part railed off, many other visitors were there as well. Fifteen dervishes came in, wrapped in brown cloaks, and with tall, brimless hats on, and sat down on the floor. A hadji in green sat upon a carpet opposite to them, and recited verses from the Koran in a droning voice, the fifteen dervishes giving occasional bows. Then a voice in a gallery struck up, and some tinkling music sounded. This was the signal for the hadji to get up, followed by the dervishes, who pulled off their brown cloaks, and let down their long, white skirts, and slowly marching round they bowed each time as they came to the bit of carpet. After this the twirling began. Holding up one hand with the palm, the other with the back uppermost, and shutting their eyes, they went round and round the place like a set of tops, with their skirts standing well out, the hadji walking in and out amongst them. We

left them spinning, and drove down to the shore, going off to the ship in caiques. We had a very merry dinner on board in the evening, Captains Grenfell and Jolliffe dining
6th May. with us. On the following day we weighed anchor at 11.30 A.M. and went up the Bosphorus into the Black Sea as far as the lightship—a lovely trip! We got back to Constantinople at 5 P.M., and having got all that was wanted from shore, started on our way across the Sea of Marmora, and passed Chanak
7th May. 8th May. again on the 7th, and anchored at the Piræus on the 8th, finding several men-of-war, principally Russian, with some Greek and French, but only one English gunboat, the *Bittern*.
9th and 10th May. The next two days were spent in visiting Athens and its environs, and on the 10th our good dragoman, Louis Mansour, who had accompanied us all the way from Cairo, left us for his home in Beyrout, to the regret of all. A more trustworthy, obliging, and excellent fellow could not be found. He had been of the greatest service and comfort to us all. We wrote our names at the palace, and received a polite message that the King would receive us if we wished, but felt we had no claim to his time, so decided not to bestow our tediousness on him.

11th May. On the morning of the 11th the *Monarch* and *Bacchante* came in. A very pretty sight, the King going off in a steam launch, and all the ships, English, French, Russian, and Greek, manning yards, and dressing ship, whilst the *Iris* saluted. This was repeated when the King left for shore, taking his nephews with him. The weather turned very disagreeable, blowing hard, but Mr. Ford, our Minister, and Lord and Lady Dufferin came on board, and spent a couple of hours.

12th May. We left Piræus at 11 A.M., and steamed slowly to the Bay of Salamis, and found the *Monarch* at anchor. Going on board her in the afternoon, we found Captain Fairfax had gone to Athens, but we had a long chat with Commander Hammil, who returned our visit later in the day, bringing a very minute midshipman with him, and some mail-bags for us to deliver at Malta.

At 5 A.M. next day we left our anchorage and had fine, pleasant weather during the voyage to Malta, where we arrived at 8 A.M on the 16th, and made fast to a buoy in Sleima Inlet, Valetta. Here we found Harry, who had come out from England to join us and finish the cruise with us. **13th to 16th May.**

We had a day's sight-seeing on the 17th, visiting the armoury of the palace, and the tapestry in the council chamber. Calling on Mr. Hoare he kindly joined us, and we went to see St. John's Church, and then to tea with Sir Victor and Lady Houlton, after which we looked in at Mr. Hoare's, and went back to dine on board. **17th May.**

To-day we all started off in carriages, and drove to San Antonio Palace, which has beautiful gardens thickly planted with lemons and oranges, Japanese medlars, and other fruit-trees; and then on to Citta Vechia, to see the old Roman villa lately discovered, which is in excellent preservation; and to the Church of St. Paul, and the catacombs. From here we went to the Verdala Palace, which is the summer residence of the governor, and occupies a fine position on a ridge of hills, from which a very perfect view is got of the greater portion of the island, and the blue Mediterranean. We lunched in the hall of the palace, and drove back to Valetta well pleased with our day, and remained at the hotel, as coaling was going on. **18th May.**

We got back to our ship to-day; the young people went **19th May.**

off to see some sports, and during their absence we had several visitors, amongst others Admiral Graham, Sir Victor Houlton, Colonel Murray, Captain and Mrs. Borton, and Mr. Hoare. It was blowing rather hard, and the ladies didn't like the coming and going in the boat.

20th May. We had more visitors again to-day, Lieutenant Onslow, the admiral's flag lieutenant, came, he had been at Coquimbo in the *Penguin* last year, but had left hurriedly to look after the *Doterel*. Then Mr. Layard, whose brother we had seen in Yokohama. Captain Simpson, the governor's A.D.C., also lunched with us. We left for Syracuse at 4 P.M., our departure being watched from the ramparts by Mr. Hoare, who had been most kind, and had advised our going to Sicily, telling us what was really worth visiting. We reached

21st May. Syracuse at 7 A.M. in the morning. The bay is long, and the vivid green of the country was a most pleasant contrast to the glaring whiteness of Malta. The town with its remains of past splendour looks well from the sea, and Etna was in full view. We went ashore for a couple of hours, and saw the fountain of Arethusa, an old church which once formed part of the temple of Minerva, the remains of the temple of Diana, a Greek and Roman amphitheatre with the seats hewn out of the rock, and Dionysius' Ear, which is a huge cave with an extraordinary echo. The story goes that it was used by Dionysius as a political prison, from the top of which he could hear the least whisper of his prisoners. Then into catacombs, where the guide told us we might wander for weeks without coming to the end of them. They originally formed the town of Syracuse, but after the introduction of Christianity

22nd May. were turned into burial grounds. We got on board and were off again by 2 P.M., and arrived at Catania at 6 P.M., and anchored just inside a new pier which is being constructed to enlarge the little mercantile harbour at present existing. We had a most delightful day, going ashore in the morning and taking the 10 A.M. train for Giardini. The country, one large orchard of peaches, pomegranates, lemons, vineyards, figs, and other fruits. We arrived at Giardini about one o'clock, and then got into carriages to go on to Taormina. We could see the town far above us, to which we zig-zagged up a steep but very good road, with long views the whole way, and ruins of temples constantly to be seen. We found a delightful little hotel, where we had lunch, and afterwards went to see the remains of a Greek theatre, and up to the signal post from whence a glorious view was obtained. The sea blue and glittering beneath us, the steep hills dotted with little towns, the valleys and flats perfect gardens, Etna close to us with a mantle of snow on one side, and a continuous volume of smoke pouring from the crater, little bays and promontories, and nearly every town defended by an old castle perched upon some commanding crag. Above Taormina we saw a town balanced on the crest of a hill where there didn't seem a level spot in the place. The guide, Valerio, whom we had brought with us from Syracuse, told us that the children as soon as they begin to walk are tied up until they are four or five years old, for fear of their falling down the hill-side. We came away from this beautiful place most reluctantly; it has a special charm of its own, and was one of the very few places we saw which made one say, "I should like to come here again."

## CHAPTER XXIII.

*PALERMO.—NAPLES.—ROME.—CAGLIARI.—ALGIERS.—GRENADA.—GIBRALTAR.—QUEENSTOWN AND COWES.*

23rd May. AT 6 A.M. on a bright, beautiful morning, we left Catania, and had a most enjoyable day's journey. As we steamed up the Straits of Messina, we had the beautiful coast scenery of Sicily on one side, that of Italy on the other, and the volcanoes of Etna and Stromboli sending up their columns of smoke. Passing between Scylla and Charybdis, we looked in at Milazzo in the evening, and found the people very busy with the tunny fishery, but didn't stay long, wanting to get

24th May. on to Palermo, where we anchored at 6 A.M. in the morning. The bay bears the poetic name of "The Golden Shell," and is certainly deserving of admiration, but the surrounding hills are dry and arid, which takes off from its beauty. As it was the Queen's birthday, we dressed ship with masthead flags, our example being speedily followed by several merchant steamers in harbour. After breakfast, we went ashore to see the Palace, Cathedral, and Museum, and found the streets good and clean, with beautiful shops and large houses, and numbers of soldiers about. In the afternoon we drove out to Monreale, a town a few miles from Palermo, situated on the side of a hill and having a very magnificent church, splendidly decorated from floor to roof with mosaics illustrating scenes

from the Old and New Testaments, and with a rich ceiling. The view over town and bay, and valleys filled with orange and other trees, was perfect, but the readers of this journal must be getting very tired, I fear, of beautiful views. We got back to the hotel Trinacria to dine, going on board to sleep. At 11 A.M. we weighed anchor and proceeded under 25th May. steam and sail with lovely weather, and before dark had lost sight of Sicily, our visit to which we owed to Mr. Hoare's well-deserved recommendation of its beauties.

Lovely weather continued throughout the night, and by 10 26th May. A.M. we were in sight of Naples, passing between the island of Capri and the mainland, and came to anchor off the Mole at 2 P.M. The Orient Company's steamer *Austral* was lying here, a very fine steamer with 1,300 passengers on board. One of her officers kindly invited us to visit the ship, which we did, and found her a magnificent vessel, fitted with every luxury and convenience.

The next four days we devoted to visiting the environs 27th, 28th, 29th, 30th May. of Naples, Pompeii and Herculaneum, and the many points of interest in Naples, all of which have been so often described that there is nothing left for me to say about them. On the last day of the month we had a very pleasant 31st May. trip in the cutter, in tow of the steam lunch, our consul, Mr. Grant, coming with us and in the most good-natured way acting as our cicerone. Skirting along the coast of the gulf of Pozzuoli, and passing between the island of Nisida and the mainland, to the island of Procida, and visiting the very interesting ruins along the coast, we landed at Baiæ to see the temples of Venus and Mercury, which stand on either side of the little hill. In the temple of Mercury, some girls danced the "Tarantella" for us, very gracefully and prettily, which reminded us somewhat of the sama-cueca of Chili and Peru. After lunch we started again, steaming close along the coast past Pauli and Mirabella, into Port Miseno, and round Cape Miseno to Procida.

The ruins along the coast between Baiæ and Cape Miseno are numerous, and in the clear water you see remains of old buildings and great masses of masonry jutting out

S. BARTOLOMEO, ROME.

like rocks beneath you. From Procida we returned to the ship, arriving about 7 P.M. after a most enjoyable day.

1st June. In the afternoon we left Naples and steamed between Procida and the mainland. Vesuvius was very active, a column of smoke pouring from the crater. About two in the afternoon

2nd June. of the next day, we anchored in the cramped little harbour of Civita Vecchia, and started for Rome by the 5.15 train.

The country was beautifully green and fresh, and the yellow broom and wild roses were in full bloom. The hay was being cut and carried, and as it grew dark, thousands of fire-flies sparkled in the hedges and fields. We put up at the hotel Costanzi, where we were very comfortable, and found very few people staying at this time of the year, and during the next few days were busily engaged in visiting the wonders and beauties of Rome, so often described. At 2 P.M. on the

3rd to 9th June.

CAGLIARI.

9th we left Rome with great regret, feeling that we had seen so little of this vast museum of antiquities, historical associations and triumphs of art, but the real lovers and students of Rome will know far better than we can, how very small a portion of its beauties and charms could have been realised in the space of a week. Arriving at Civita Vecchia at 4 P.M., we started for Cagliari at 5 P.M., meeting a stiffish breeze and hard sea, which increased to heavy squalls during the night.

10th June. The morning was however light and lovely, both wind and sea fast going down as we approached the island of Sardinia and after a fine night we anchored in Cagliari harbour at
11th June. 9 A.M. It is an open roadstead, and the wind blew again in heavy squalls, preventing our landing until towards evening, when we took a stroll through the town and up to the citadel, and arranged for a visit next day to the Montevecchio mines, which are very extensive and produce large quantities of argentiferous lead.

Next day we landed early, leaving the ladies on board,
12th June. and left by the first train at 6 A.M. for Montevecchio, passing through a broken country, with plains covered with olive trees and corn. We ascended the mountain range in which the mines are situated by the company's tramway as far as the first mine, and then got into carriages to ascend to the summit, where the manager's house is situated, and where we were most kindly entertained at breakfast. We then descended on the opposite side of the mountain to visit another large mine belonging to the same company, and then took train again for Cagliari, and
13th and 14th June. were on board and on our way to Algiers by 9 P.M.; where we arrived at 5.30 P.M. on the 14th. Algiers and its environs have lately been so charmingly described by Mr. Knox in his *New Playground*, that I will only say we were not disappointed with it, and that having walked and driven about and seen as much as we could in so short a time, we
15th June. left for Malaga in the afternoon, and had a heavy thunderstorm
16th June. during the night. In the afternoon, just before sunset, we met the Channel squadron under steam and sail, coming along in

two lines, the first ship signalled asking our name, and giving hers in return, *Minotaur;* then came the *Agincourt* and *Northumberland*, followed by the *Sultan* and *Achilles* in the other line. It was an imposing sight, and we watched them fade away in the distance on their voyage to the East.

Next day we sighted land at about 2 A.M., and having anchored outside Malaga harbour at eight in the evening, we burned some blue lights to attract the attention of a pilot, who, when he came off, told us that we had done wisely in not trying to enter the cramped little harbour. Early on the following day he shifted us a little nearer to the mouth of the harbour, but we decided not to enter, and thus avoided the trouble of mooring head and stern, as well as the horrible smells. It was Sunday, so we had our usual service, and after luncheon took a drive through the town and visited the English cemetery, and the beautiful residence of Mr. Heredia. After paying several visits in the morning, we left by the midday train for Grenada, reaching the town about 8 P.M. after a very hot journey. Next morning we woke to a charming scene. It was a lovely morning, with a soft and cool breeze; on all sides sounded the trickling and murmuring of water. We breakfasted out of doors, under shady trees, with fountains and flowers round us. The pomegranate trees were lovely, with their bright red flowers amidst the green leaves. The day was spent in and about the Alhambra and the Generalife Palace. The latter place now belongs to Count Palavicini, who married the owner of the property, and whose gardens near Genoa must be known to many English tourists. We were shown round by the gardener's daughter, a beautiful girl with splendid eyes, but

17th June.

18th June.

19th June.

20th June.

unfortunately too conscious of her charms. She has ma up her mind to marry none of lower rank than a prince, sh says, so I expect she will continue to show admiring crow over the Generalife for some time to come. From the gard you can trace numbers of holes in the neighbouring hil sides, inhabited by hundreds of gipsies. The king of th gipsies, a handsome fellow, was successfully sketched b Mr. Pritchett.

I shall not attempt any description of the wonder and beauties of this romantic place, with its relics of Moorish grandeur, or of its views over the Vega to the snow-
21st June. capped mountains of the Sierra Nevada. This day we spen according to the fancy of the various members of our party visits being paid to the palace of Charles V., the tomb of Ferdinand and Isabella, the Carthusian convent with it lovely marbles and cabinets, to Mr. Contrera's studio, wher some of his beautiful casts were bought, and Senor Calderon's
22nd June. house and garden. We made an early start, being called at 3 A.M., and after breakfast left the hotel at half-past four for the station. Such a lovely morning! The nightingales singing, and the morning light shining through the branches of the trees in the avenues. We got back to Malaga at 1 P.M. and went on board rather tired, and glad, as we always were, to get back to the ship, after two or three days ashore. We
23rd June. had many visitors to-day. Unfortunately, although the sea was smooth there was a good deal of motion, and I fear that all the ladies did not enjoy themselves as much as we could have wished.

24th June. Next morning at 4 A.M. we started for Gibraltar, arriving at 2 P.M, after a pleasant run of ten hours, and got a capital berth inside the New Mole, through the kindness of Captain

Fremantle. We were busy paying visits the next two days, 25th and 26th June.
and again on the 27th, when we began the day by break- 27th June.
fasting on board the American yacht, *Namouna*, belonging to Mr. Gordon Bennett, meeting the governor, Lord Napier of Magdala, and Lady Napier, and some friends of Mr. Bennett's who were cruising about with him. After breakfast Lord and Lady Napier, and his aide-de-camp, came on board the *Wanderer* with us, and afterwards we went in Captain Fremantle's steam launch to visit the *Hecla* torpedo ship, possibly the ugliest and the most deadly ship in Her Majesty's service. Before we left the Reserve Squadron hove in sight, seven ironclads coming into harbour, under the command of the Duke of Edinburgh, in line of battle. The sight was a grand one, but it had a special interest for us, as one of the ships was under the command of our old friend Captain Cator. After they had got into place I went on board the *Lord Warden* and brought back Cator to dinner, after which Bob, Cator, and I attended a levée on board the *Hercules*, being most graciously received by the Duke of Edinburgh, and finding the Duke of Connaught also on board, who had come for a little change of air.

Two of our men were overheard discussing these two Royal dukes, and one of them said: "Could we find out whether they send any of their half pay to their widdy mother?" To-
day being Coronation Day there was a heavy expenditure of 28th June.
gunpowder from batteries on shore and men-of-war in the bay. The saluting was deafening, with echoes rumbling and
rolling round the rock. The Governor had kindly sent us a 29th June.
card of invitation to dinner, but I was not able to go myself, so Captain Cator escorted Mrs. Lambert. There was a large party, with the Dukes of Edinburgh and Connaught present,

after which Lady Napier was "at home" in the garden, wh
30th June. was illuminated with Chinese lanterns. We spent the d
chiefly in calling on the captains of the ships composing t
1st July. Reserve Squadron, and next day had visitors at 5 o'cl
tea, and in the evening I dined on board the *Hercules* w

DRAGON TREE IN THE GOVERNOR'S GARDEN, GIBRALTAR.

the Duke of Edinburgh. The Detached Squadron came in during the day, and they seemed quite like old friends, this being the fourth time we had met them: at Fiji, Hong
2nd July. Kong, Singapore, and now at Gibraltar. We had our Sunday service as usual in the morning, and were busy in the afternoon receiving the Dukes of Edinburgh and Connaught.

who honoured us with a visit. We also had visitors from shore, and from many of the captains of the Reserve and Detached Squadrons, and from Lord and Lady Napier, so that it was a very gay afternoon.

At 6.30 A.M. Nellie and Beatrice went ashore and rode with Lord and Lady Napier up to the O'Hara tower, Lady Napier and Miss Napier coming back to breakfast with us on board. There was to be a grand field day, sailors and marines to land, the Duke of Edinburgh to be saluted by the guns in the galleries, and all sorts of nautical and martial manœuvres to take place, but we had made up our minds to leave to-day, so at noon we were steaming away from the midst of it all, and I found myself wondering how I could have had resolution enough to leave the many kind friends, old as well as new, and to decline all the generous hospitalities that were offered on all sides, both ashore and afloat. Indeed, our gratitude was felt, but could not be worthily expressed to Lord and Lady Napier, Major Gilbard, and many others to whom we owed so much that made the visit to Gibraltar the most memorable of all our voyage. Steaming along we passed the Pillars of Hercules, and were out in the open at 1 P.M. on the 4th, and entered the Atlantic, and after a breezy night with considerable head sea we anchored in the river off Lisbon at 7 A.M., and having received our letters went on our way again at 3 P.M. Next day we got a favourable breeze, continuing for the next two days with occasional squalls of rain with lightning.

3rd July. 4th July. 5th July. 6th July. 7th and 8th July.

This morning at 4 A.M. we sighted land, and at 8 A.M. were shackled to a man-of-war buoy in Queenstown harbour, and were very soon greeted by the welcome faces of our daughter Kate and her husband Captain Clutterbuck, to be followed

9th July.

by many other relations, who had come over from Eng... to meet us here. It was Sunday, and we had our serv... 10.30 A.M.

We stayed in Queenstown until the 15th, when ... left for Milford, the weather during our stay being v... boisterous and disagreeable.

16th July. On the 16th, as we were entering Milford harbour we w... met by the *Kelpie* yacht, belonging to my brother-in-... and crowded with nephews and nieces innumerable, who h... in spite of very inclement weather, come out to give u...
17th July. hearty welcome home. It is needless to say what a hap... meeting it was, or what a merry day was spent. Cap... Vine very kindly took us over Pembroke dockyard to-d...
18th July. and at 4 A.M. next morning we started for our final de... tination at Cowes, the sea right in our teeth, and the sh... pitching heavily. Between 8 and 9 P.M. we sighted th... Wolf Light, and at 10 P.M. were able to shift our course up Channel, and so turn the foul wind into a fair one, and with the aid of our canvas skim along at the rate of from ten to
19th July. twelve knots an hour. It was lovely weather, and at 1 P.M. we began to shorten and furl sails, having St. Alban's Head abeam. By 4 P.M. we were ahead of Yarmouth, and shortly after in sight of R.Y.S. Castle, from whence "welcome home again" was signalled, and at 5 P.M. we were anchored in our old berth, with the familiar spots and scenes around us, after an absence of nearly two years.

Two years of singular blessings to ourselves, preserved as we had been, thank God, from all dangers of land and sea, and in which we had stored up a fund of pleasant memories to serve us for many a future year. Memories not alone of new and beautiful scenes, but of kind faces and warm hearts. Alas! that the grateful recognition of these new friendships only made us feel more keenly the absence of the old friends for whose hearty welcome home we had looked with such delight, foremost amongst whom would have been the friend whose brotherly sympathy had been extended to me and mine through so many years of a somewhat chequered life.

R.Y.S. "WANDERER."

*Abstract from Log-book of "Wanderer,"*

| 1880. | | | Temperature of Watr. Deg. Fahr. | Air Deg. Fahr. Ht. | Lt. | Position at Noon. Latitude. Deg. | Min. | Longitude. Deg. | Min. | Course. |
|---|---|---|---|---|---|---|---|---|---|---|
| Aug. 5 | Leave Cowes ... ... ... ... | 7·15 P.M. | 53 | 62 | 70 | ... | ... | ... | ... | Various |
| „ 6 | High hd. sea; hosd. tpmats. & Tt. Yd. | ... | 55 | 74 | 64 | 49 | 15 N. | 4 | 47 W. | W.S.W. |
| „ 7 | High beam sea; ship rolling heavily | ... | 64 | 65 | 63 | 46 | 58 | 7 | 7 | S.W. |
| „ 8 | Weather fining; sea going down; sight Cape P. ... ... ... ... | ... | 65 | 67 | 64 | 43 | 50 | 8 | 37 | S.W. |
| „ 9 | Arrive at Vigo ... ... ... ... | 6·30 A.M. | 67 | 82 | 58 | ... | ... | ... | ... | Various |
| | *Duration of voyage, 3½ days.* | | | | | | | | | |
| | TOTALS ON VOYAGE ... ... ... | ... | ... | ... | | ... | ... | ... | ... | — |
| „ 11 | Leave Vigo ... ... ... ... | 11.35 A.M. | 62 | ... | | ... | ... | ... | ... | Various |
| „ 12 | Arrive at Lisbon... ... ... ... | 11.35 A.M. | 63·44 | 71 | 66 | ... | ... | ... | ... | .. |
| | *Duration of voyage, 1 day.* | | | | | | | | | |
| „ 21 | Leave Lisbon ... ... ... ... | 12 noon | 63·67 | ... | | ... | ... | ... | ... | Various |
| „ 22 | ... ... | ... | 69·70 | 78 | 69 | 36 | 58 N. | 11 | 12 W. | W.S.W. |
| „ 23 | ... ... | ... | 74·70 | 78 | 70 | 34 | 42 | 13 | 45 | — |
| „ 24 | ... ... | ... | ... | 78 | 69 | 33 | 29 | 15 | 20 | W. by S. ½ S. |
| „ 25 | Arrive at Funchal ... ... ... | 12 noon | 75 | 76 | 69 | ... | ... | ... | ... | Various |
| | *Duration of voyage, 4 days.* | | | | | | | | | |
| | TOTALS ON VOYAGE ... ... ... | ... | ... | ... | | ... | ... | ... | ... | — |
| „ 27 | Leave Funchal ... ... ... ... | 11 A.M. | 79 | 79 | 74 | 32 | 33 N | 16 | 31 W. | S. by E. |
| „ 28 | Feathered screw. Peak of Teneriffe in sight ... ... ... ... | ... | ... | 80 | 73 | 29 | 48 | 16 | 22 | S. by W. |
| „ 29 | ... ... | ... | ... | 78 | 73 | 27 | 33 | 16 | 51 | Various |
| „ 30 | ... ... | ... | ... | 75 | 83 | 25 | 16 | 18 | 22 | S. 31° 0 W. |
| „ 31 | ... ... | ... | ... | 86 | 74 | 23 | 7 | 20 | 7 | S. 36° 29′ W. |
| Sept. 1 | ... ... | ... | ... | 85 | 76 | 20 | 0 | 22 | 7 | S. W. by W. ¼ W. |
| „ 2 | ... ... | ... | ... | 85 | 75 | 18 | 28 | 24 | 36 | S. 45° W. |
| „ 3 | ... ... | ... | 79 | 84 | 78 | 16 | 38 | 25 | 44 | Various |
| „ 4 | Began to steam at ... ... ... | 9 A.M. | 80 | 86 | 79 | 14 | 35 | 25 | 46 | S. |
| „ 5 | ... ... | ... | 80·81 | 79 | 81 | 12 | 47 | 23 | 33 | S. 46° E. |
| „ 6 | ... ... | ... | 81 | 84 | 79 | 10 | 48 | 21 | 25 | 45 |
| „ 7 | ... ... | ... | 80·79 | 85 | 79 | 9 | 30 | 18 | 52 | 63 |
| „ 8 | ... ... | ... | 79·78 | 80 | 77 | 8 | 2 | 15 | 41 | 65 W |
| „ 9 | ... ... | ... | 79·80 | 79 | 77 | 6 | 47 | 12 | 54 | 66 |
| „ 10 | ... ... | ... | 79·80 | 79 | 75 | 5 | 41 | 10 | 17 | 70 |
| „ 11 | ... ... | ... | 79·80 | 79 | 76 | 3 | 40 | 7 | 23 | 51 |
| „ 12 | ... ... | ... | 79 | 79 | 74 | 3 | 12 | 4 | 12 | 67 |
| „ 13 | ... ... | ... | 74·75 | 78 | 75 | 2 | 34 | 1 | 11 | 74 |
| „ 14 | ... ... | ... | 75·77 | 80 | 77 | 2 | 21 | 1 | 23 E. | 76 |
| „ 15 | ... ... | ... | 76·77 | 81 | 75 | 1 | 32 | 3 | 36 | 70 |
| „ 16 | ... ... | ... | 77 | 80 | 77 | 0 | 45 | 6 | 29 | 76 |
| „ 17 | Arrive at Gaboon ... ... ... | 2 P.M. | 79 | 79 | 76 | ... | ... | .. | ... | Various |
| | *Duration of voyage, 21½ days.* | | | | | | | | | |
| | TOTALS ON VOYAGE ... ... ... | ... | ... | ... | | ... | ... | ... | ... | — |
| „ 26 | Leave Gaboon ... ... ... ... | 6.30 A.M. | 79 | 82 | 81 | 0 | 8 N | 9 | 3 E. | Various |
| „ 27 | ... ... | ... | 79 | 82 | 77 | 1 | 17 S. | 6 | 23 | S. 62° W. |
| „ 28 | ... ... | ... | 76·79 | 81 | 76 | 3 | 27 | 4 | 52 | Various |
| „ 29 | ... ... | ... | 76·74 | 79 | 74 | 6 | 15 | 3 | 43 | S. 21′ 55 W. |
| „ 30 | ... ... | ... | 78·71 | 76 | 72 | 9 | 1 | 2 | 35 | 22 |
| Oct. 1 | ... ... | ... | 72 | 71 | 69 | 11 | 0 | 0 | 15 | 49 30′ |
| „ 2 | ... ... | ... | 70 | 78 | 67 | 13 | 9 | 1 | 58 W. | 45 13 |
| „ 3 | ... ... | ... | 70 | 67 | 65 | 15 | 16 | 4 | 56 | 46 30 |
| „ 4 | Arrive at St. Helena ... ... ... | 6.30 A.M. | 70 | 68 | 66 | ... | ... | ... | ... | Various |
| | *Duration of voyage, 8 days.* | | | | | | | | | |
| | TOTALS ON VOYAGE ... ... ... | ... | ... | ... | | ... | ... | ... | ... | — |
| „ 7 | Leave St. Helena... ... ... ... | 10.35 A.M. | 68 | 64 | 78 | ... | ... | ... | ... | S. 83° W. |
| „ 8 | ... ... | ... | ... | 65 | 78 | 15 | 48 S. | 7 | 27 W | — |
| „ 9 | ... ... | ... | ... | 66 | 71 | 15 | 44 | 9 | 37 | N. 88° W. |
| „ 10 | ... ... | ... | ... | 66 | 72 | 15 | 37 | 11 | 18 | 85° 57′ |
| „ 11 | ... ... | ... | ... | 67 | 74 | 15 | 24 | 13 | 12 | 83 |
| „ 12 | ... .. | ... | 73 | 68 | 72 | 15 | 11 | 15 | 16 | 84 1 |
| „ 13 | ... ... | ... | 73 74 | 71 | 74 | 14 | 57 | 17 | 57 | 84 41 |
| „ 14 | ... ... | ... | 74·75 | 70 | 76 | 14 | 40 | 21 | 25 | 85 19 |
| „ 15 | ... ... | ... | 75·74 | 71 | 78 | 14 | 23 | 23 | 46 | 83 |
| „ 16 | ... ... | ... | 75 | 72 | 77 | 14 | 12 | 26 | 30 | 86 |
| „ 17 | ... ... | ... | 77 | 74 | 79 | 13 | 59 | 29 | 24 | 85 |
| „ 18 | ... ... | ... | 77 | 80 | 76 | 13 | 43 | 32 | 4 | 84 |
| „ 19 | ... ... | ... | 78 | 81 | 75 | 13 | 19 | 34 | 50 | 81 |
| „ 20 | Arrive at Bahia ... ... ... ... | 4 P.M. | 78 | 82 | 79 | ... | ... | ... | ... | Various |
| | *Duration of voyage, 13 days 6 hours.* | | | | | | | | | |
| | TOTALS ON VOYAGE ... ... ... | ... | ... | ... | | ... | ... | ... | ... | — |

*n from Place of Departure to Place of Arrival.*

| ...stances run under Steam and Sail. | | | | | Average Speed per hour. | Coal used, with average per mile run under steam. | | Wind and Weather. | | Barometer. | | Remarks. |
|---|---|---|---|---|---|---|---|---|---|---|---|---|
| . | Steam and Sail. | Total Stm. | Sail. | Grand Total Run. | | Tons. | Avge. per Run. | Direction. | Force. | Hst. | Lst. | |
| . | Mls. | Mls. | Mls. | Mls. | Mls. | | Cwts. | | | | | |
| | ... | ... | ... | ... | ... | ... | ... | ... | 0 | 29 96 | | |
| ) | ... | 160 | ... | 160 | 10 | ... | ... | W.N.W. | 7 | 30·40 | 29·78 | |
| 3 | ... | 226 | ... | 226 | 9 41 | ... | ... | W .by S. | 7 | 29·87 | 29 90 | |
| 0 | ... | 200 | ... | 200 | 8 33 | ... | ... | W.S.W. | 6.3 | 29·34 | 30·22 | |
| 5 | ... | 135 | ... | 135 | 7·54 | 34 | ·913 | N. by E. | 8.0 | 30·50 | 30·12 | Stay in Vigo two days. |
| 1 | ... | 721 | ... | 721 | ... | ... | ... | ... | ... | 30·27 | 30·11 | |
| | ... | ... | ... | ... | ... | ... | ... | Calm. | ... | 30·09 | 30 02 | |
| ·7 | ... | 257 | ... | 257 | 10·70 | 16·25 | 1 264 | ... | ... | 30 | 29·98 | Stay in Lisbon nine days. |
| | ... | ... | ... | ... | ... | ... | ... | W. by N. | 0.2 | 30·06 | | |
| 50 | ... | 150 | ... | 150 | 6·25 | ... | ... | S.W. to W. by N. | 2.5 | 30·04 | 30·19 | |
| . | 177 | 177 | ... | 177 | 7·37 | ... | ... | W. by N. | 1.2.8 | 30·18 | 30·24 | |
| . | ... | ... | 114 | 114 | 4·75 | ... | ... | W.N.W. | 3.4 | 30·18 | 30·23 | |
| 30 | ... | 80 | 66 | 96 | 4 | ... | ... | W.N.W.—N. by E. | 1.8.2 | 30·20 | 30·29 | |
| 80 | 177 | 357 | 180 | 537 | 5·59 | 11·50 | ·634 | ... | ... | 30 31 | 30·34 | |
| 11 | ... | 11 | ... | 11 | 10 | ... | ... | E N.E. | 2.1 | 30·30 | 30 27 | |
| ... | ... | ... | 172 | 172 | 7·11 | ... | ... | E.N E. | 4.8.2 | 30 27 | 30·21 | |
| ... | ... | ... | 150 | 150 | 6 25 | ... | ... | N.E. | 2.3 | 30·21 | 30·15 | |
| ... | ... | ... | 160 | 160 | 6·66 | ... | ... | E.N.E. | 2.3 4 | 30·10 | 30·19 | |
| ... | ... | ... | 160 | 160 | 6 66 | ... | ... | N.E. by E. E.N.E. | 4 3.2 | 30·13 | 30·13 | |
| ... | ... | ... | 180 | 180 | 7 5 | ... | ... | E N.E. N.E. by S. | 4.2.5 | 30·10 | 30·14 | |
| ... | ... | ... | 194 | 194 | 8·08 | ... | ... | N.E. by E —E. | 2 4 4.2 | 30·10 | 30·19 | |
| ... | ... | ... | 139 | 139 | 5 78 | ... | ... | E. to E N.E. | 2.3 | 30 6 | 30·12 | |
| ... | 23 | 23 | 99 | 122 | 5 | ... | ... | S.E.—N.E.—E. | 1.2.1 | 30·7 | 30·13 | |
| ... | 182 | 182 | ... | 182 | 7 6 | ... | ... | S.E.—E. by N. nim. | 1.3 5 | 30·7 | 30·11 | |
| 173 | ... | 173 | ... | 173 | 7·20 | ... | ... | N.W. to S.E. nim. | 1 2 0 | 30·4 | 30·10 | |
| ... | 173 | 173 | ... | 173 | 7·20 | ... | ... | S.W.—S.S.W. nim. | 2.5 | 30 | 30·05 | |
| ... | 208 | 208 | ... | 208 | 8·66 | ... | ... | S.W.—W.S.W. nim. | 3.5 | 30 | 30·09 | |
| ... | 181 | 181 | ... | 181 | 7·54 | ... | ... | W S.W. nim. | 2.5 | 30·02 | 30·29 | |
| 160 | ... | 160 | ... | 160 | 6·66 | ... | ... | W.S.W. to. S nim. | 5 | 30·05 | 30·15 | |
| ... | 194 | 194 | ... | 194 | 8·08 | ... | ... | S.W. by S. nim. | 5.6 | 30·14 | 30 06 | |
| ... | 16 | 16 | 177 | 193 | 8 08 | ... | ... | S.W. to S.S.W. cirro. cu. | 2.5 | 30 | 30·12 | |
| ... | ... | ... | 184 | 184 | 7·7 | ... | ... | S. by W. | 2.5 | 30 | 30·19 | |
| ... | ... | ... | 155 | 155 | 6 45 | ... | ... | S by W. to S.S.W. cu. st | 2 | 30·01 | 30·12 | |
| ... | ... | ... | 142 | 142 | 5 91 | ... | ... | S.S.W. cu. st. & cirro. cu. | 2.4 | 30·21 | 30·05 | |
| ... | ... | ... | 200 | 200 | 8 33 | ... | ... | S.W. cirro. cu. & nim. | 4 | 30·14 | 30·04 | |
| 206 | ... | 206 | 10 | 216 | 8·30 | ... | ... | S.S.W. nim. | 5.6 | 30 02 | 30·19 | |
| 550 | 977 | 1,527 | 2,122 | 3,649 | 7·20 | 52 25 | ·681 | | | | | |
| 30 | ... | 30 | ... | 30 | 5 | ... | ... | S.S.W. cu. stratus | 4 | 30·08 | ... | |
| 181 | ... | 181 | ... | 181 | 7·54 | ... | ... | W.S.W cu. stra. | 2.4 | 30·03 | 30·12 | |
| 169 | ... | 169 | ... | 169 | 7·04 | ... | ... | S.S.W.—W. by S. nim. | 4.5 | 30·03 | 30 12 | |
| 184 | ... | 184 | ... | 184 | 7·66 | ... | ... | S.W. to S. by W. cu. | 4.5 | 30 02 | 30·11 | |
| 179 | ... | 179 | ... | 179 | 7·87 | ... | ... | S.W.—S. | 2.4 | 30·08 | 30·09 | |
| 133 | .. | 133 | 49 | 182 | 7·58 | ... | ... | S.S.W cu. & nim. | 4.5 | 30·02 | 30·13 | |
| ... | ... | ... | 183 | 183 | 7 62 | ... | ... | S. to S.S E. | 5 | 30 05 | 30·19 | |
| ... | ... | ... | 214 | 214 | 8 91 | ... | ... | S.S E. cirro. & cu. | 5.7 | 30·07 | 30·21 | |
| 30 | ... | 30 | 30 | 60 | ... | ... | ... | S.S.E. to S.E | 5 | 30·15 | 30 23 | Lying-to all night for dayit. |
| 906 | ... | 906 | 476 | 1,382 | 7·17 | 40·50 | ·894 | | | | | |
| ... | ... | ... | 8 | 8 | ... | ... | ... | S S.E. to S.E co. cu. | 4 2 4 1 | 30·12 | 30·15 | To clear the ships only. |
| ... | ... | ... | 97 | 97 | 4·04 | ... | ... | S E to S E. by E. cum. | 2.4.2 | 30·10 | 30 15 | |
| ... | ... | .. | 131 | 131 | 5·45 | ... | ... | S.E. to E S.E. ni. cu | 2 | 30·10 | 30·15 | |
| ... | ... | ... | 99 | 99 | 4 11 | ... | ... | S. to S.S.E cumulus | 2 0 | 30·10 | 30 20 | |
| ... | ... | ... | 101 | 101 | 4·20 | ... | ... | S.E. by E co cu. | 2 3 | 30·10 | 30 17 | |
| ... | ... | ... | 124 | 124 | 5 16 | ... | ... | E. by S. S.S.W. E S.E.cu. | 4 2 | 30·12 | 30 19 | |
| ... | 118 | 118 | 27 | 145 | 6·04 | ... | ... | S.S E to N W ni. cu. | 4.3 | 30·14 | 30 21 | |
| ... | 208 | 208 | ... | 208 | 8 66 | ... | ... | S E by E.—E.S.E co. cu. | 2.4.2 | 30·16 | 30·22 | |
| ... | 120 | 120 | 14 | 134 | 5 60 | ... | ... | E.S.E co. cu. & str. | 4.2.5 | 30·15 | 30·23 | |
| ... | ... | ... | 163 | 163 | 6·80 | ... | ... | E.S.E co. str. | 4 | 30·10 | 30 19 | |
| | ... | ... | 163 | 162 | 6 80 | ... | ... | E.S.E. cumulus. | 4 | 30·10 | 30·17 | |
| ... | ... | ... | 161 | 161 | 6 70 | ... | ... | E. by S.—N E. co. cu. | 4.5 | 30·10 | 30 15 | |
| ... | ... | ... | 161 | 161 | 6 70 | ... | ... | N.E. by N. cumulus. | 4 | 30 06 | 30·16 | |
| 210 | ... | 210 | 14 | 224 | 8 | ... | ... | Various. cu. cir. str. | 0.2 | 30·10 | 30·17 | |
| 210 | 446 | 656 | 1,263 | 1,919 | 6·03 | 28·25 | ·861 | E. by S —E.S.E. | 4.2 | 30 09 | 30·15 | |

*Abstract from Log-book of "Wanderer,"*

| 1880. | | | Temperature of | | | Position at Noon. | | | | Course. |
|---|---|---|---|---|---|---|---|---|---|---|
| | | | Watr. | Air. | | Latitude. | | Longitude. | | |
| | | | Deg. Fahr. | Deg. Fahr. Ht. | Lt. | Deg. | Min. | Deg. | Min. | |
| Oct. 26 | Leave Bahia ... ... ... ... | 5 P.M. | 80 | 80 | 75 | ... | ... | ... | ... | Various |
| ,, 27 | ... ... | ... | 78 | 80 | 73 | 15 | 16 S. | 38 | 11 W | — |
| ,, 28 | High sea; ship pitching heavily ... | ... | 76 | 77 | 75 | 17 | 10 | 38 | 14 | S. 1° E. |
| ,, 29 | ,, ,, rolling ... ... ... | ... | 75 | 82 | 74 | 20 | 6 | 38 | 30 | [illegible] W |
| ,, 30 | ... ... | ... | 72 | 83 | 73 | 22 | 44 | 40 | 59 | 42° |
| ,, 31 | Arrive at Rio Janeiro ... ... ... Lying-to for daylight. | 8 A.M. | 70 | 74 | 86 | ... | ... | ... | ... | Various |
| | *Duration of voyage, 4 days 15 hours.* | | | | | | | | | |
| | TOTALS ON VOYAGE ... ... ... | ... | ... | ... | | ... | ... | ... | ... | — |
| Nov. 13 | Leave Rio ... ... ... ... ... | 6 P.M. | ... | ... | | ... | ... | ... | ... | — |
| ,, 14 | Calm and light winds ... ... .. | ... | 75 | 81 | 72 | 25 | 10 S. | 44 | 34 W. | S. 31 W |
| ,, 15 | Fine night and nice fair breeze; all sqr. sails set ... ... ... ... | ... | 73 | 80 | 74 | 28 | 10 | 46 | 51 | 34 |
| ,, 16 | Heavy head sea ... ... ... ... | ... | 78 | 75 | 69 | 30 | 26 | 48 | 48 | 36 |
| ,, 17 | ... ... | ... | 67 | 68 | 62 | 32 | 7 | 49 | 50 | 28 |
| ,, 18 | Strong breeze and high head sea ... | ... | 61 | 62 | 57 | 33 | 21 | 51 | 27 | 49 |
| ,, 19 | Light breeze; sea diminishing ... | ... | 64 | 62 | 55 | 35 | 18 | 54 | 46 | 58 |
| ,, 19 | Arrive at Monte Video ... ... ... | 8 P.M. | ... | 73 | 59 | ... | ... | ... | ... | Various |
| | *Duration of voyage, 6 days.* | | | | | | | | | |
| | TOTALS ON VOYAGE ... ... ... | ... | ... | ... | | ... | ... | ... | ... | — |
| ,, 22 | Leave Monte Video ... ... ... | 8 P.M. | 65 | 76 | 68 | ... | ... | ... | ... | Various |
| ,, 23 | Arrive at Buenos Ayres ... ... | 7 A.M. | 69 | ... | | ... | ... | ... | ... | — |
| | *Duration of voyage, 15 hours.* | | | | | ... | ... | ... | ... | |
| ,, 30 | Leave Buenos Ayres ... ... ... | 5 P.M. | 67 | 71 | 60 | ... | ... | ... | ... | — |
| Dec. 1 | Arrive at Monte Video ... ... ... | 7 A.M. | 65 | 66 | 60 | ... | ... | ... | ... | — |
| | *Duration of voyage, 14 hours.* | | | | | | | | | |
| ,, 4 | Leave Monte Video ... ... ... | 5 P.M. | 68 | 77 | 62 | ... | ... | ... | ... | — |
| ,, 5 | Lovely day ... ... ... ... | ... | 68·61 | 76 | 67 | 37 | 31 S. | 56 | 30 W | — |
| ,, 6 | Heavy N.E. swell ... ... ... | ... | 57 | 65 | 60 | 39 | 54 | 59 | 2 | S. 40 W |
| ,, 7 | Fine and clear ... ... ... ... | ... | 56 | 68 | 61 | 41 | 50 | 61 | 53 | 48 |
| ,, 8 | Arrive at Pyramid Bay, New Gulf ... | 10 A.M. | 56 | 72 | 61 | ... | ... | ... | ... | Various |
| ,, 9 | Run across Gulf to Port Madryn ... Sailed 6 A.M., arrived 10 P.M. | ... | 60 | 79 | 62 | ... | ... | ... | ... | — |
| ,, 10 | Heavy gale from N.E. on dead lee shore ... ... ... ... | ... | 59 | ... | | ... | ... | ... | ... | — |
| ,, 11 | Return and arrive at Port Madryn ... | 9 A.M. | 57·60 | 70 | 68 | ... | ... | ... | ... | — |
| | *Time under way, 14 hours.* | | | | | | | | | |
| ,, 11 | Leave Port Madryn again ... ... | 4 15 P.M. | 57·60 | ... | | ... | ... | ... | ... | — |
| ,, 12 | Anchor off Chuput river ... ... | 6.15 A.M. | 57 | 64 | 59 | ... | ... | ... | ... | Various |
| ,, 13 | Leave Chuput ... ... ... ... | 5 A.M. | 56 | 57 | 61 | 44 | 27 | 64 | 50 | S. 5° E |
| ,, 14 | ... ... ... ... | ... | 50 | 61 | 55 | 47 | 26 | 65 | 10 | 5 W |
| ,, 15 | ... ... ... ... | ... | 46 | 59 | 58 | 51 | 01 | 66 | 53 | 18 |
| ,, 16 | Anchor under Condor Cliff ... ... | 6.30 A.M. | 46 | 54 | 44 | ... | ... | ... | ... | Various |
| | *Duration of voyage, 8 days.* | | | | | | | | | |
| | TOTALS ON VOYAGE ... ... ... | ... | ... | ... | | ... | ... | ... | ... | ... |
| ,, 19 | Leave Cape Virgins ... ... ... | 2.30 A.M. | ... | ... | | ... | ... | ... | ... | — |
| | Anchor at Sandy Point ... ... | 4.30 P.M. | 46 | 43 | 54 | ... | ... | ... | ... | Various |
| ,, 22 | Leave Sandy Point ... ... ... | 10 50 A.M. | 46 | 44 | 54 | ... | ... | ... | ... | ... |
| | Anchor off Elizabeth Island ... | 2.50 P.M. | | | | ... | ... | ... | ... | ... |
| ,, 27 | Leave Elizabeth Island ... ... | 6 A.M. | 46 | 56 | 44 | ... | ... | ... | ... | — |
| | Anchor at Sandy Point ... ... | 9 P.M. | | | | | | | | |
| ,, 30 | Leave Sandy Point ... ... ... | 4.10 A.M. | 47 | 61 | 47 | ... | ... | ... | ... | — |
| | Anchor at Borja Bay ... ... | 7.10 P.M. | | | | | | | | |
| ,, 31 | Leave Borja Bay... ... ... ... | 4.15 A.M. | 48 | 44 | 55 | ... | ... | ... | ... | .. |
| 1881 | Anchor Coal Mine, Skyring Water | 5.45 P.M. | | | | | | | | |
| Jan. 1 | Leave Skyring Water ... ... ... | 4 A.M. | 50 | 50 | 45 | ... | ... | ... | ... | — |
| | Havannah Point ... ... ... | 8 P.M. | | | | | | | | |
| ,, 2 | From off Havannah Point ... ... | 6 A.M. | 48 | 45 | 52 | ... | ... | ... | ... | — |
| | Anchor at Isthmus Bay ... ... | 3.30 P.M. | | | | | | | | |
| ,, 3 | Leave Isthmus Bay ... ... ... | 6.30 A.M. | 49 | 47 | 57 | ... | ... | ... | ... | ... |
| | Anchor at Puerto Bueno ... ... | 6 30 P.M. | | | | | | | | |
| ,, 5 | Leave Puerto Bueno ... ... ... | 4.30 A.M. | 49 | 57 | 50 | ... | ... | ... | ... | ... |
| | Anchor at Port Grappler ... ... | 7 P.M. | | | | | | | | |
| ,, 6 | Leave Port Grappler ... ... ... | 8.30 A.M. | 50 | 46 | 51 | ... | ... | ... | ... | ... |
| | Return to Port Grappler, going round Saumarez Island ... ... ... | 12.15 P.M. | | | | | | | | |
| ,, 7 | Leave Port Grappler ... ... ... | 5 A.M. | 49 54 | 51 | 40 | ... | ... | ... | ... | ... |
| | Sombrero Islands abeam ... ... | 8 P.M. | | | | | | | | |
| | *Time under way, 5 days 4½ hours.* | | | | | | | | | |
| | TOTALS IN STRAITS ... ... ... | ... | ... | ... | | ... | ... | ... | ... | ... |

*from Place of Departure to Place of Arrival.*

| ...nces run under Steam and Sail. | | | | Average Speed per hour. | Coal used, with average per mile run under steam. | | Wind and Weather. | | Barometer. | | Remarks. |
|---|---|---|---|---|---|---|---|---|---|---|---|
| ...team and Sail. | Total Stm. | Sail. | Grand Total run. | | Tons. | Avge per Run. | Direction. | Force. | Hst. | Lst. | |
| Mls. | Mls. | Mls. | Mls. | Mls. | | Cwts | | | | | |
| ... | ... | ... | ... | ... | ... | ... | E. by S.—S.S.E. cum. | 4·2 | 30·17 | 30·24 | |
| 185 | 185 | ... | 185 | 7·01 | ... | ... | S.E. by S —S.S.E. nim. | 2·5 | 30 21 | 30·25 | |
| 120 | 120 | ... | 120 | ·5 | ... | ... | S.E. nim | 5·7 | 30·28 | 30 32 | |
| 176 | 176 | ... | 176 | 7 88 | ... | ... | S.E.—E. cum. str. | 6·7 | 30·25 | 30 31 | |
| 209 | 209 | ... | 209 | 8·70 | ... | ... | E.N.E.—N E. cirro. cu. | 4 6 | 30 24 | 30·1[illegible] | |
| ... | 118 | ... | 118 | 5·65 | ... | ... | N.E. | 0·2 | 29·99 | 29·91 | Very heavy thunderstorm and floods of rain. |
| 640 | 758 | ... | 758 | 6 78 | 28·50 | ·624 | | | | | |
| ... | ... | ... | ... | ... | ... | ... | ... | ... | 30 | ... | A glorious evening & night. |
| ... | 156 | ... | 156 | 8·66 | ... | ... | S.W. by S. cirro. stra. | 0·2 | 30 | 30·04 | |
| 218 | 218 | ... | 218 | 9·06 | ... | ... | S. by E. to N. by E. nim. | 2·4 | 30·02 | 29·91 | |
| 170 | 170 | ... | 170 | 7·08 | ... | ... | N. to S.W. by S. nim. | 6 8 | 29 76 | 30·06 | |
| 115 | 115 | ... | 115 | 4·80 | ... | ... | S. by W. to S.W cu. st. ci. | 4·6 | 30·15 | 30·25 | |
| ... | 113 | ... | 113 | 4·70 | ... | ... | S.W. by S. cirrus. | 4 6 | 30·22 | 30·27 | |
| 119 | 119 | ... | 119 | 8·30 | ... | ... | S.W. to N.W. cirro. cu. | 2 5 | 30 20 | 30 12 | |
| ... | 74 | ... | 74 | 9·25 | ... | ... | N. | 2·0 | 30 05 | 29·96 | |
| 702 | 1,045 | ... | 1,045 | 7·15 | 41 | ·669 | | | | | |
| ... | ... | ... | ... | ... | ... | ... | N W.—N.N.W. cirr. | 4 | 30·15 | 30·10 | |
| ... | 115 | ... | 115 | 7 66 | 7·75 | 1·343 | N.W.—N.N.W. cum. | 4·6 | 30 07 | 30·04 | |
| ... | ... | ... | ... | ... | ... | ... | E. by N. to N. cirro. cu. | 2·7 | 30·28 | 30 25 | |
| ... | 115 | ... | 115 | 8 21 | 7·25 | 1 260 | N. cirro. str. | 4·7 | 30 25 | 30 26 | |
| ... | ... | ... | ... | ... | ... | ... | W. cumulus | 2·0 | 29·82 | 29·87 | |
| ... | 160 | ... | 160 | 8·42 | ... | ... | S.W.—S.S E. cir. str. | 0 2 | 29·80 | 29 99 | |
| 186 | 186 | ... | 186 | 7·75 | ... | ... | W. cir. str. | 2·5 | 29·97 | 29·80 | |
| 178 | 178 | ... | 173 | 7·20 | ... | ... | S.W.—N.E cir. cu. | 2·5 | 29·86 | 29 60 | |
| 133 | 133 | ... | 133 | 6 | ... | ... | W.—W N.W. cir. cu. | 0·7 | 29·49 | 29·84 | Slowed down for daylight. |
| ... | 85 | ... | 85 | 8·75 | ... | ... | E.N.E. cir. str. | 4·5 | 29·82 | 30·20 | |
| ... | ... | ... | ... | ... | ... | ... | ... | ... | 30·10 | 29·85 | Got under way 7 P.M., wind veering round compass. Steamed dead slow across gulf & back with S.W. wd. |
| ... | 60 | ... | 60 | 4·30 | ... | ... | Round compass | 7·8 | 29·59 | 29·89 | |
| ... | ... | ... | ... | ... | ... | ... | N.E. cirro. cu. | 5 | 29·85 | 30·0 | |
| ... | 75 | ... | 75 | 5 85 | ... | ... | S.W. to E. cirro. cu | 5·7 | 29 94 | 29 64 | |
| 59 | 59 | ... | 59 | 8 43 | ... | ... | W.S.W. cumulus | 2·6 | 29·70 | 29 90 | |
| 180 | 180 | ... | 180 | 7 50 | ... | ... | S.—N.W. cir. str. | 6·2 | 29·94 | 29·89 | |
| 225 | 225 | ... | 225 | 9 87 | ... | ... | N.W.—W.S.W. cum. | 4·6 | 29 70 | 29·06 | |
| ... | 108 | ... | 108 | 5·79 | ... | ... | W.S.W. cum. | 8·9 | 28 95 | 28·84 | |
| 956 | 1,394 | ... | 1,394 | 7·18 | 58·25 | ·764 | Furious gale, S.W. | 8·9 | 28·98 | 29·04 | Furious gale from W S.W. to S.W.; ride it out 16th, 17th, 18th. Lowest glass, 28·65. |
| ... | ... | ... | ... | ... | ... | ... | cirro. cu. | 5 6 | ... | ... | Stiff S.W. breeze. |
| ... | 124 | ... | 124 | 7·75 | ... | ... | S.W. cirro. cum. | 4·6 | 29·35 | 29 47 | |
| ... | 24 | ... | 24 | 6 | ... | ... | W. cumulus | 2 5 | 29 74 | 29·96 | |
| ... | 24 | ... | 24 | 8 | ... | ... | N.W.—N.E. cu. cir. cu. | 5 | 29·62 | 29·45 | |
| ... | 104½ | ... | 104½ | 6·96 | ... | ... | N. to W. | 2 | 29·61 | 29 34 | |
| ... | 91 | ... | 91 | 6 74 | ... | ... | W. to N.W. cu. nim. | 2·5 | 29·32 | 29·42 | Aiding *Santos*. |
| ... | 127½ | ... | 127½ | 8 | ... | ... | W.—N.W. nim. | 0·4 | 29·42 | 29·52 | Stmg. slow off & on all night. |
| ... | 85½ | ... | 85½ | 9 | ... | ... | N.W. nim. cu. | 5·2 | 29·48 | 29 91 | Very thick and raining hard. |
| ... | 82 | ... | 82 | 6·83 | ... | ... | N.W. cumulus | 5 | 29·94 | 29 72 | |
| ... | 111½ | ... | 111½ | 7 67 | ... | ... | ... | 6·2 | 29·42 | 29·71 | |
| ... | 46 | ... | 46 | 8 | ... | ... | nim. | 2·7 | 29·71 | 29·43 | Down Chasm Reach, returning by Tay Reach. |
| ... | 97½ | ... | 97½ | 6·50 | ... | ... | ... | 5·7 | 29·37 | 29·74 | |
| ... | 917 | ... | 917 | 7·38 | | | | | | | |

*Abstract from Log-book of "Wan…*

| 1881. | | | Temperature of Watr. Deg. Fahr. | Air. Deg. Fahr. Ht. | Lt. | Position at Noon. Latitude. Deg. | Min. | Longitude. Deg. | Min. | C… |
|---|---|---|---|---|---|---|---|---|---|---|
| Jan. 8 | From Sombrero yesterday at ... | 8 P.M. | 52·57 | 49 | 59 | 46 | 41 S. | 76 | 23W | Va… |
| " 9 | ... ... ... | ... | 55·60 | 55 | 64 | 43 | 23 | 75 | 51 | N… |
| " 10 | ... ... ... | ... | 60 62 | 62 | 59 | 39 | 54 | 74 | 38 | N… |
| " 11 | ... ... ... | ... | 62 | 60 | 65 | 36 | 22 | 74 | 03 | N… |
| " 12 | Made fast to buoy Valparaiso ... | 1 P.M. | 56 | 59 | 72 | — | — | — | — | Va… |
| | Left Valparaiso ... ... ... | 6 P.M. | 59·61 | | | | | | | |
| " 13 | Anchor in Coquimbo ... ... | 3 P.M. | 59·61 | 61 | 70 | — | — | — | — | Va… |
| | *Duration of voyage, 5 days 14 hrs.* | | | | | | | | | |
| | Totals on Voyage ... ... ... | ... | ... | ... | | — | ... | — | — | – |

WORK IN…

| Date | | | Watr. | Air Ht. | Lt. | Lat. Deg. | Min. | Long. Deg. | Min. | C… |
|---|---|---|---|---|---|---|---|---|---|---|
| Feb. 15 | Left Coquimbo ... ... ... | Noon | 61 | | | | | | | |
| " 16 | Arrived at Valparaiso ... ... | 4 45 P.M. | 60 63 | ... | | — | — | — | — | S… |
| Mar. 2 | Left " ... ... ... | 10.30 A.M. | 58 60 | ... | | — | ... | — | — | N… |
| " 3 | Arrived at Tongoy ... ... ... | 9 A.M. | 59·62 | ... | | — | ... | — | — | – |
| " 3 | Left " ... ... ... | 9 30 P.M. | | | | | | | | |
| " 8 | Arrived at Coquimbo ... ... | 3.40 P.M. | ... | ... | | — | — | — | — | Va… |
| " 26 | Left " ... ... ... | Noon | 62 58 | | | | | | | |
| " 27 | Arrived at Valparaiso ... ... | 6 30 P.M. | 60 59 | ... | | — | — | — | — | – |
| April 3 | Left " ... ... ... | 6 30 A.M. | 60 61 | | | | | | | |
| " 4 | Returned to " ... ... ... | 8.30 A.M. | 60 | | | | | | | |
| " 11 | Left " ... ... ... | 3.45 P.M. | 61 | ... | | ... | — | — | — | – |
| " 12 | Returned to " ... ... ... | 10.15 A.M. | 60,61,62 | ... | | ... | — | — | — | – |
| " 13 | Left " ... ... ... | 4 P.M. | 62 | 61 | 62 | ... | — | ... | — | Va… |
| " 14 | Arrived at Coquimbo ... ... | Noon | 62 | 65 | 62 | ... | — | ... | — | – |
| May 14 | Left " ... ... .. | 10 A.M. | 57,58,59 | 61 | 70 | — | — | — | — | Va… |
| " 15 | Arrived at Valparaiso ... ... | 10 A.M. | 59 | 58 | 63 | | | | | |
| " 16 | Left " ... ... ... | Noon | 59 | 58 | 66 | — | — | — | — | Va… |
| " 17 | At sea ... ... ... ... ... | ... | 59 | 61 | 60 | 32 | 58 S. | 74 | 38 W. | N… |
| " 18 | " ... ... ... ... | ... | 60·61 | 57 | 60 | 32 | 50 | 76 | 41 | N… |
| " 19 | Arrived at Cumberland Bay, Juan Fernandez ... ... ... ... | 4 P.M. | 60,61,62 | 61 | 68 | ... | ... | ... | — | S… |
| " 20 | Left " " ... ... | Noon | 61 | 59 | 61 | ... | ... | — | — | – |
| " 21 | At sea ... ... ... ... ... | ... | 60·61 | 63 | 57 | 33 | 19 | 77 | 49 | Var N… |
| " 22 | " ... ... ... ... | ... | 61·59 | 58 | 63 | 33 | 12 | 75 | 34 | N… |
| " 23 | " ... ... ... ... | ... | 59 | 62 | 64 | 33 | 4 | 73 | 20 | N… |
| " 24 | Arrived at Valparaiso ... ... | 1 P.M. | 59·58 | 61 | 64 | ... | ... | ... | ... | Va… |
| " 27 | Left " ... ... ... | 9.20 P.M. | ... | ... | | ... | ... | ... | ... | – |
| " 29 | Returned to " ... ... ... | 2.10 P.M. | 59 63 | 57 | | ... | ... | ... | ... | Va… |
| " 31 | Left " ... ... ... | 4 P.M. | 57·59 | 56 | 58 | ... | ... | ... | — | – |
| June 1 | Arrived at Coquimbo ... ... | 1 P.M. | 60·63 | 58 | 57 | ... | ... | ... | ... | Va… |
| " 7 | Left " ... ... ... | 7 P.M. | 55·57 | 58 | | | | | | |
| " 8 | Arrived at Valparaiso ... .. | 3.30 P.M. | 59·62 | 58,59,60 | | ... | ... | ... | ... | Va… |
| " 8 | Left " ... ... ... | 4 30 P.M. | 61·65 | ... | | | | | | |
| " 9 | Arrived at Coquimbo ... ... | 2.30 P.M. | 62·67 | 50,60,57 | | ... | ... | ... | ... | Va… |
| | *Time under way, 19 days 22 hours.* | | | | | | | | | |
| | Totals of Navigation in Chili ... | ... | ... | ... | | ... | ... | — | ... | – |

VOYAGE CONT…

| Date | | | Watr. | Air Ht. | Lt. | Lat. Deg. | Min. | Long. Deg. | Min. | C… |
|---|---|---|---|---|---|---|---|---|---|---|
| June 16 | Leave Coquimbo ... ... ... | 4 P.M. | 60·59 | 56 | 59 | ... | ... | ... | ... | – |
| " 17 | At sea ... ... ... ... ... | Noon | 59·60 | 59 | | 30 | 37 S. | 73 | 16W | N… |
| " 18 | " ... ... ... ... | ... | 58·56 | 61 | | 30 | 37 | 77 | 10 | N… |
| " 19 | " ... ... ... ... | ... | 60·57 | 61 | | 30 | 11 | 80 | 59 | N… |
| " 20 | " ... ... ... ... | ... | 64·58 | 63 | | 29 | 49 | 84 | 27 | N… |
| " 21 | " ... ... ... ... | ... | 60·58 | 63 | | 29 | 43 | 86 | 40 | N… |
| " 22 | " ... ... ... ... | ... | 63·58 | 63 | 64 | 29 | 27 | 88 | 19 | N… |
| " 23 | Very high beam sea, southerly ... | ... | 68 57 | 66 | | 29 | 8 | 91 | 29 | N… |
| " 24 | " " " ... | ... | 65·60 | 66 | | 27 | 59 | 93 | 35 | N… |
| " 25 | " " " ... | ... | 61·71 | 66 | | 26 | 13 | 94 | 33 | N… |
| " 26 | " " " ... | ... | 65·71 | 68 | | 25 | 0 | 96 | 18 | N… |
| " 27 | Swell moderated ... ... ... | ... | 65·71 | 68 | | 24 | 7 | 97 | 52 | N… |
| " 28 | Swell greatly reduced ... ... | ... | 65·76 | 70 | | 23 | 8 | 99 | 58 | N… |
| " 29 | Swell quite lost ... ... ... | ... | 67·78 | 71 | | 22 | 16 | 102 | 40 | N… |
| " 30 | " ... ... ... ... | ... | 67·78 | 71 | | 21 | 22 | 105 | 33 | N… |
| | Carried Forward ... ... | ... | ... | ... | | ... | ... | — | — | – |

*. from Place of Departure to Place of Arrival.*

| ances run under Steam and Sail. | | | | Average Speed per hour. | Coal used, with average per mile run under steam. | | Wind and Weather. | | Remarks. | | |
|---|---|---|---|---|---|---|---|---|---|---|---|
| | | | | | | | | | Barometer | | |
| Steam and Sail. | Total Stm. | Sail. | Grand Total Run. | | Tons. | Avge. per Run. | Direction. | Force | Hst. | Lst. | |
| Mls. | Mls | Mls. | Mls | Mls. | | Cwts | | | | | |
| ... | 66 | ... | 66 | 4 12 | ... | ... | S.W.—N.W. cir. cu. | 5 | 29·79 | 29·83 | Very high sea. |
| 200 | 200 | ... | 200 | 8·33 | ... | ... | N E. cumulus | 4·2 | 29 64 | 29·77 | Long swell from S.E. . . . |
| 216 | 216 | ... | 216 | 9 | ... | ... | W.N.W. cir. str. nim. | 2·4 | 29·82 | 30 08 | |
| 213 | 213 | ... | 213 | 8·87 | ... | ... | N.W.—W.S.W. cu. | 2·6 | 30·11 | 30·24 | |
| ... | 222 | ... | 222 | 8·88 | ... | ... | S. cir. str. | 2·6 | 30·21 | 30·01 | |
| ... | 180 | ... | 180 | 8·77 | ... | ... | Calm cir. cu. | 0 | 30·03 | 30·17 | Long swell S. |
| 629 | 1,097 | ... | 1,097 | 8·18 | 113·50 | 1·127 | | | | | |

STAY IN CHILI.

| Steam and Sail. | Total Stm. | Sail. | Grand Total Run. | Average Speed per hour. | Tons. | Avge. per Run. | Direction. | Force | Barometer | Remarks. |
|---|---|---|---|---|---|---|---|---|---|---|
| ... | 200 | ... | 200 | 6·89 | ... | ... | S. | 2·4 | ... | Heavy head sea. |
| 180 | 180 | ... | 180 | 8 | ... | ... | S.W. | 5 | ... | Very fine. |
| ... | 21 | ... | 21 | 7 | ... | ... | Calm | 0 | ... | Very fine. |
| ... | 200 | ... | 200 | 5 33 | ... | ... | S.W. | 5·6 | ... | Fires banked while in Port. High head sea. |
| 50 | 50 | ... | 50 | 2 | ... | ... | N.W. gale | 7 | ... | Steam and try-sails. |
| ... | ... | ... | ... | ... | ... | ... | N.W. gale | 6·7 | ... | Steam and try-sails. |
| 40 | 40 | ... | 40 | 2 15 | ... | ... | N.W. gale | ... | ... | Dirty and heavy rain. |
| ... | ... | ... | ... | ... | ... | ... | ... | ... | ... | Fine and clear. |
| 200 | 200 | ... | 200 | 10 | ... | ... | S.W.—S.E.—N.E. | 2 | ... | Lovely weather. |
| ... | 200 | ... | 200 | 8·33 | ... | ... | S. to W. | 3 | ... | Fine. |
| ... | 16 | 135 | 151 | 6·29 | ... | ... | S.S.E.—S.—S.W. | 4·2 | ... | Moderate, fine, squalls. |
| ... | ... | 96 | 96 | 4 | ... | ... | S. and variable. | 2 | ... | Ditto, and strong N. currt. |
| 113 | 113 | 6 | 119 | 4·25 | ... | ... | S.W. to S.E. | 2 | ... | Ditto, with showers. |
| ... | ... | ... | ... | ... | ... | ... | ... | ... | ... | Two hrs. lying off and on in English Bay, while party vistd. Alex. Selkirk's cave. |
| 12 | 12 | 56 | 68 | 3 | ... | ... | S. to calm. | 0·3 | ... | Fine, cloudy. |
| 9 | 91 | 12 | 103 | 4·3 | ... | ... | S.W. to S.S.E. | 0·2 | ... | Fine, cloudy. |
| 103 | 103 | ... | 103 | 4·8 | ... | ... | W. to calm | 2 | ... | Stormy. |
| 103 | 108 | ... | 103 | 4·10 | ... | ... | W. to calm | 1 | ... | Stormy. |
| ... | ... | ... | ... | ... | ... | ... | ... | ... | ... | Put to sea for a norther. |
| 50 | 50 | ... | 50 | ... | ... | ... | N. to N.W. | 8·2 | ... | Heavy sea, rain. |
| ... | ... | ... | ... | ... | ... | ... | N. to calm. | 2 | ... | Very fine. |
| 200 | 200 | ... | 200 | 9·52 | ... | ... | N. to E. | 2·5 | ... | Very fine. |
| 200 | 200 | ... | 200 | 9·75 | ... | ... | S.W.—N.—N.W. | 2 | ... | Very fine. |
| ... | 200 | ... | 200 | 9·09 | ... | ... | N to N.E. | 2 | ... | Very fine. |
| 1,342 | 2,179 | 305 | 2,484 | 5·19 | 109·3 | 1 | | | | |

.TH SEA ISLANDS.

| Steam and Sail. | Total Stm. | Sail. | Grand Total Run. | Average Speed per hour. | Tons. | Avge. per Run. | Direction. | Force | Barometer | Remarks. |
|---|---|---|---|---|---|---|---|---|---|---|
| ... | ... | ... | ... | ... | ... | ... | ... | ... | Cumulus | |
| ... | 114 | ... | 114 | 5 70 | ... | ... | S. to calm. | 2 | Very fine | Very fine. |
| 52 | 52 | 149 | 201 | 8·37 | ... | ... | S.E. to S. by W. | 5·6 | ... | Fresh and clear. |
| ... | ... | 200 | 200 | 8·33 | ... | ... | S.S.E. | 5 6 | Cirr. cu. | Fresh and squally. |
| ... | ... | 185 | 185 | 7·70 | ... | ... | S.S.E. | 6·4 | Cirr cu. | Fresh to moderate. |
| ... | ... | 115 | 115 | 4·79 | ... | ... | S.E.—E.—W. by S. | 5·2 | Cumulus | Light breeze, cloudy. |
| ... | 67 | 18 | 85 | 3·54 | ... | ... | W. by S. | 2 7 | Cumulus | Light breeze, calm, cloudy. |
| ... | 25 | 148 | 173 | 7·20 | ... | ... | W by S. to S.E. | 2·7 | Cirro. cu. | Light to stiff breeze, squally. |
| ... | 15 | 115 | 130 | 5·41 | ... | ... | E. to S.S.E —W. | 2 | Cumulus | Light, moderate, fine. |
| 91 | 91 | 27 | 118 | 4·91 | ... | ... | W. by N to S.S.W. | 8·4 | Cumulus | Very heavy squalls to fine. |
| ... | ... | 120 | 120 | 5 | ... | ... | S. to S.E. | 4·2 | Cumulus | Fine, moderate, squally. |
| ... | ... | 101 | 101 | 4·20 | ... | ... | S.E. to E N.E. | 4·2 | Cumulus | Fine, moderate, squally. |
| ... | ... | 127 | 127 | 5,3,0 | ... | ... | E. by S. | 2·4 | Cumulus | Very fine. |
| ... | ... | 162 | 162 | 6·75 | ... | ... | E. by S. to S. E. | 4·2 | Cirro. cu. | Very fine. |
| ... | ... | 167 | 167 | ·7 | ... | ... | E. to E. by N. | 5·2 | Cirro. cu. | Fine, with rain, squally. |
| 143 | 361 | 1,634 | 1,998 | 6·01 | | | | | | |

*Abstract from Log-book of "Wander[illegible]"*

| 1881. | | | Temperature of Watr. Deg. Fahr. | Temperature of Air. Deg. Fahr. (Ht. Lt.) | Position at Noon. Latitude. Deg. | Latitude. Min. | Longitude. Deg. | Longitude. Min. | Course. |
|---|---|---|---|---|---|---|---|---|---|
| Jan. 8 | From Sombrero yesterday at ... | 8 P.M. | 52·57 | 49 59 | 46 | 41 S. | 76 | 23 W | Vari[illegible] |
| ,, 9 | ... ... ... | ... | 55·60 | 55 64 | 43 | 23 | 75 | 51 | N. 7[illegible] E |
| ,, 10 | ... ... ... | ... | 60·62 | 62 59 | 39 | 54 | 74 | 38 | N. 1[illegible] E |
| ,, 11 | ... ... ... | ... | 62 | 60 68 | 36 | 22 | 74 | 03 | N. [illegible] E |
| ,, 12 | Made fast to buoy Valparaiso ... | 1 P.M. | 56 | 59 72 | ... | ... | ... | ... | Vari[illegible] |
| | Left Valparaiso ... ... ... | 6 P.M. | 59·61 | | | | | | |
| ,, 13 | Anchor in Coquimbo ... ... | 3 P.M. | 59·61 | 61 70 | ... | ... | ... | ... | Vari[illegible] |
| | *Duration of voyage, 5 days 14 hrs.* | | | | | | | | |
| | Totals on Voyage ... ... ... | ... | ... | ... | ... | ... | ... | ... | — |

WORK DONE [illegible]

| 1881. | | | Watr. | Air. | Lat. Deg. | Lat. Min. | Long. Deg. | Long. Min. | Course. |
|---|---|---|---|---|---|---|---|---|---|
| Feb. 15 | Left Coquimbo ... ... ... | Noon | 64 | | | | | | |
| ,, 16 | Arrived at Valparaiso ... ... | 4 45 P.M. | 60 63 | ... | ... | ... | ... | ... | Southw[illegible] |
| Mar. 2 | Left ,, ... ... ... | 10.30 A.M. | 58·60 | ... | ... | ... | ... | ... | North[illegible] |
| ,, 3 | Arrived at Tongoy ... ... ... | 9 A.M. | 59·62 | ... | ... | ... | ... | ... | — |
| ,, 3 | Left ,, ... ... ... | 9 30 P.M. | | | | | | | |
| ,, 3 | Arrived at Coquimbo ... ... | 3.40 P.M. | ... | ... | ... | ... | ... | ... | Vari[illegible] |
| ,, 26 | Left ,, ... ... ... | Noon | 62 58 | | | | | | |
| ,, 27 | Arrived at Valparaiso ... ... | 8 30 P.M. | 60 59 | ... | ... | ... | ... | ... | — |
| April 3 | Left ,, ... ... ... | 6 30 A.M. | 60 61 | | | | | | |
| ,, 4 | Returned to ,, ... ... ... | 8 30 A.M. | 60 | | | | | | |
| ,, 11 | Left ,, ... ... ... | 3.45 P.M. | 61 | ... | ... | ... | ... | ... | — |
| ,, 12 | Returned to ,, ... ... ... | 10.15 A.M. | 60,61,62 | ... | ... | ... | ... | ... | — |
| ,, 13 | Left ,, ... ... ... | 4 P.M. | 62 | 61 62 | ... | ... | ... | ... | Vari[illegible] |
| ,, 14 | Arrived at Coquimbo ... ... | Noon | 62 | 65 62 | ... | ... | ... | ... | — |
| May 14 | Left ,, ... ... .. | 10 A.M. | 57,58,59 | 61 70 | ... | ... | ... | ... | Vari[illegible] |
| ,, 15 | Arrived at Valparaiso ... ... | 10 A.M. | 59 | 58 63 | ... | ... | ... | ... | Vari[illegible] |
| ,, 16 | Left ,, ... ... ... | Noon | 59 | 58 66 | | | | | |
| ,, 17 | At sea ... ... ... ... ... | ... | 59 | 61 60 | 32 | 58 S. | 74 | 38 W. | N. 69 W |
| ,, 18 | ,, ... ... ... ... | ... | 60·61 | 57 60 | 32 | 50 | 76 | 41 | N. 86 W |
| ,, 19 | Arrived at Cumberland Bay, Juan Fernandez ... ... ... ... | 4 P.M. | 60,61,62 | 61 63 | ... | ... | ... | ... | S. 79 W |
| ,, 20 | Left ,, ,, ... ... | Noon | 61 | 59 61 | ... | ... | ... | ... | — |
| ,, 21 | At sea ... ... ... ... ... | ... | 60·61 | 63 57 | 33 | 19 | 77 | 40 | Var. N. [illegible] E |
| ,, 22 | ,, ... ... ... ... | ... | 61·59 | 58 63 | 33 | 12 | 75 | 34 | N. [illegible] E |
| ,, 23 | ,, ... ... ... ... | ... | 59 | 62 64 | 33 | 4 | 73 | 20 | N. [illegible] E |
| ,, 24 | Arrived at Valparaiso ... ... | 1 P.M. | 59·58 | 61 64 | ... | ... | ... | ... | Vari[illegible] |
| ,, 27 | Left ,, ... ... ... | 9.20 P.M. | ... | ... | ... | ... | ... | ... | — |
| ,, 29 | Returned to ,, ... ... ... | 2.10 P.M. | 59 63 | 57 | ... | ... | ... | ... | Vari[illegible] |
| ,, 31 | Left ,, ... ... ... | 4 P.M. | 57·59 | 56 58 | ... | ... | ... | ... | — |
| June 1 | Arrived at Coquimbo ... ... | 1 P.M. | 60·63 | 58 57 | ... | ... | ... | ... | Vari[illegible] |
| ,, 7 | Left ,, ... ... ... | 7 P.M. | 55·57 | 58 | | | | | |
| ,, 8 | Arrived at Valparaiso ... ... | 3.30 P.M. | 59 62 | 58,59,60 | ... | ... | ... | ... | Vari[illegible] |
| ,, 8 | Left ,, ... ... ... | 4.30 P.M. | 61·65 | ... | | | | | |
| ,, 9 | Arrived at Coquimbo ... ... | 2.30 P.M. | 62 67 | 59,60,57 | ... | ... | ... | ... | Vari[illegible] |
| | *Time under way, 19 days 22 hours.* | | | | | | | | |
| | Totals of Navigation in Chili ... | ... | ... | ... | ... | ... | ... | ... | — |

VOYAGE CONTIN[illegible]

| 1881. | | | Watr. | Air. | Lat. Deg. | Lat. Min. | Long. Deg. | Long. Min. | Course. |
|---|---|---|---|---|---|---|---|---|---|
| June 16 | Leave Coquimbo ... ... ... | 4 P.M. | 60·59 | 56 59 | ... | ... | ... | ... | — |
| ,, 17 | At sea ... ... ... ... ... | Noon | 59 60 | 59 | 30 | 37 S. | 73 | 16 W | N. [illegible] W |
| ,, 18 | ,, ... ... ... ... | ... | 58·56 | 61 | 30 | 37 | 77 | 10 | West |
| ,, 19 | ,, ... ... ... ... | ... | 60·57 | 61 | 30 | 11 | 80 | 59 | N. [illegible] W |
| ,, 20 | ,, ... ... ... ... | ... | 64·58 | 63 | 29 | 49 | 84 | 27 | N. [illegible] W. |
| ,, 21 | ,, ... ... ... ... | ... | 60·58 | 63 | 29 | 43 | 86 | 40 | N. [illegible] W |
| ,, 22 | ,, ... ... ... ... | ... | 63 58 | 63 64 | 29 | 27 | 88 | 19 | N. [illegible] W |
| ,, 23 | Very high beam sea, southerly ... | ... | 63·57 | 66 | 29 | 8 | 91 | 29 | N. [illegible] W |
| ,, 24 | ,, ,, ,, ... | ... | 65·60 | 66 | 27 | 59 | 93 | 05 | N. [illegible] W |
| ,, 25 | ,, ,, ,, ... | ... | 61·71 | 66 | 26 | 13 | 94 | 33 | N. [illegible] W |
| ,, 26 | ,, ,, ,, ... | ... | 65·71 | 68 | 25 | 0 | 96 | 18 | N. 32 W |
| ,, 27 | Swell moderated ... ... ... | ... | 65 71 | 68 | 24 | 7 | 97 | 52 | N. [illegible] W |
| ,, 28 | Swell greatly reduced ... ... | ... | 65·76 | 70 | 23 | 8 | 99 | 58 | N. 62 W |
| ,, 29 | Swell quite lost ... ... ... | ... | 67·78 | 71 | 22 | 16 | 102 | 40 | N. 71 W. |
| ,, 30 | ,, ... ... ... ... | ... | 67 78 | 71 | 21 | 22 | 105 | 33 | N. 71 W |
| | Carried Forward ... ... | ... | ... | ... | ... | ... | ... | ... | — |

*...en fro...*

...istances ...

...R STAY IN ...

...OUTH SEA ISLANDS.

...alls.
...in sqls.
...lls.
...y light
...to clear.
...r.
...in sqls.
...water.
...squalls.
...E S. fine.
...rain sqls.
...ne.
...es.
... & en-
... Ros-
... sqls.
... sqls.
... s and
... d in
... osch-
... which

*Abstract from Log-book of "Wanderer," R.Y.*

| 1881. | | | Temperature of | | | Position at Noon. | | | | Course. |
|---|---|---|---|---|---|---|---|---|---|---|
| | | | Air. | | Watr. | Latitude. | | Longitude | | |
| | | | Deg. Fahr. | | Deg. Fahr. | Deg. | Min. | Deg. | Min. | |
| | | | Ht | Lt. | | | | | | |
| | Brought Forward ... ... | ... | ... | | ... | ... | ... | ... | ... | ... |
| July 1 | At sea ... ... ... ... ... | Noon | 69 | 78 | 72 | 20 | 37 S. | 108 | 7 W | N. 72 W. |
| " 2 | " ... ... ... ... ... | ... | 69 | 76 | 72 | 19 | 47 | 111 | 4 | 73 |
| " 3 | " ... ... ... ... ... | ... | 73 | 79 | 73 | 19 | 11 | 115 | 8 | 73 |
| " 4 | " ... ... ... ... ... | ... | 72 | 78 | 74 | 18 | 32 | 114 | 58 | N. 70 |
| " 5 | " (30·15—30·24) ... ... | ... | 74 | 82 | 74 | 17 | 39 | 117 | 22 | 69 |
| " 6 | " ... ... ... ... | ... | 75 | 80 | 75 | 16 | 52 | 119 | 55 | 72 |
| " 7 | " ... ... ... ... | ... | 75 | 81 | 74 | 16 | 8 | 112 | 36 | 74 |
| " 8 | " (30·13—30·20) ... ... | ... | 74 | 80 | 76 | 15 | 27 | 125 | 22 | 75 |
| " 9 | " (30·12—30·19) ... ... | ... | 76 | 82 | 76½ | 14 | 40 | 127 | 27 | 69 |
| " 10 | " (30 11—30·19) ... ... | ... | 78 | 84 | 78 | 14 | 16 | 128 | 32 | 69 |
| " 11 | Commenced steaming (30·4—30 19) | 2 A.M. | 78 | 84 | 79·78 | 13 | 37 | 130 | 32 | 71 |
| " 12 | " (30 12—30·21) ... ... | ... | 76 | 82 | 79 | 12 | 46 | 133 | 02 | 70 |
| " 13 | " (30·10—30·15) ... ... | ... | 79 | 82 | 79 | 11 | 51 | 136 | 2 | 73 |
| | " (30 10—30·15) | | | | | | | | | |
| " 14 | Anchor at Bon Repos Bay, Fatou Hiva, Marquesas Islands ... | 2.55 P.M. | 79 | 85 | 80 | 10 | 42 | 138 | 30 | 65 |
| | *Duration of voyage, 27 days 23 hours.* | | | | | | | | | |
| | Totals on Voyage ... ... ... | ... | ... | | ... | ... | ... | ... | ... | — |

| 1881. | | | Air Ht | Air Lt | Watr. | Lat. Deg. | Lat. Min. | Long. Deg. | Long. Min. | Course. |
|---|---|---|---|---|---|---|---|---|---|---|
| July 16 | Leave Fatou-Hiva (30·10—15) ... | 6 A.M. | 78 | 85 | 81·80 | 10 | 2 | 138 | 57 | Various |
| " " | Arrive at Resolution Bay, Taou-ata (30·10—30·15) ... ... ... | 3.50 P.M. | 78 | 85 | ... | ... | ... | ... | ... | Various |
| " " | Leave Resolution Bay (30·10 - 30·15) | 5.30 P.M. | ... | | ... | ... | ... | ... | ... | — |
| " 17 | Arrive at Hakachu Bay, Nouka-Hiva (30·05—30·16) ... ... | 8.40 A.M. | 80 | 85 | 80 | ... | ... | ... | ... | Various |
| " 19 | Leave Hakachu Bay (30·05—30·11) | 10.30 A.M. | 79 | 88 | 80 | 9 | 05 | 140 | 04 | Various |
| " " | Arrive at Hakahetau and steam down Western coast to Obelisk Island, leaving it (30·08—30·10) 5.45 P.M. | ... | 79 | 82 | ... | ... | ... | ... | ... | Various |
| | Totals of Voyages from Island to Island and Duration of Passages | | | | | | | ... | ... | — |

| 1881. | | | Air Ht | Air Lt | Watr. | Lat. Deg. | Lat. Min. | Long. Deg. | Long. Min. | Course. |
|---|---|---|---|---|---|---|---|---|---|---|
| July 19 | Leave Obelisk Island, Southern end of Roa-Poua at ... ... ... | 5·45 P.M. | ... | | ... | ... | ... | ... | ... | — |
| " 20 | At sea (30·04—30 10) ... ... | Noon | 79 | 82 | 80 | 10 | 36 S. | 141 | 48 W | S. 55 W |
| " 21 | " (30·05—30·13) ... ... | ... | 75 | 82 | 80 | 12 | 11 | 143 | 46 | 50 |
| " 22 | " (30·05—30·13) ... ... | ... | 78 | 82 | 81·80 | 13 | 58 | 146 | 1 | 51 |
| " 23 | " (30·04—30·14) ... ... | ... | 78 | 82 | 80 | 14 | 31 | 146 | 32 | Various |
| | Land at Ahu Atoll, Paumoto, 2 hrs., left 11.30 P.M. | | | | | | | | | |
| " 24 | " (30·07—30·16) ... ... | ... | 78 | 84 | 80 | 17 | 15 | 149 | 5 | Various |
| " " | Anchor in Papeete, Society Group | 4.30 P.M. | ... | | ... | ... | ... | ... | ... | — |
| | *Duration of voyage, 4 days 22½ hours.* | | | | | | | | | |
| | Totals on Voyage ... ... ... | ... | ... | | ... | ... | ... | ... | ... | — |

| 1881. | | | Air Ht | Air Lt | Watr. | Lat. Deg. | Lat. Min. | Long. Deg. | Long. Min. | Course. |
|---|---|---|---|---|---|---|---|---|---|---|
| Aug. 6 | Leave Papeete, Tahiti (30·20) ... | 11 A.M. | 84 | 82 | 78 | ... | ... | ... | ... | — |
| | Arrive at Taloo Harbour, Moorea | 1.45 P.M. | | | | | | | | |
| " 8 | Leave Taloo, Moorea (30·14—30·19) | 4 30 P.M. | 71 | 82 | 79 | ... | ... | ... | ... | — |
| | Arr. at Utuvoa H., Raiatea, 9th | 10 A.M. | | | | | | | | — |
| " 12 | Leave Utuvoa H., Raiatea (30·07) ... | 7.30 A.M. | 71 | 42 | 79 | ... | ... | ... | ... | — |
| | Arr. at Oteavanua H., Bora-bora | | | | | | | | | |
| | *Duration of Passages, 1 day 1½ hrs.* | | | | | | | | | |
| | Totals of Voyages from Island to Island ... ... ... ... | ... | ... | | ... | ... | ... | ... | ... | — |

*·ken from Place of Departure to Place of Arrival.*

| Distances run under Steam and Sail. | | | | | Average Speed per hour. | Coal used with average per mile run under steam. | | Wind and Weather. | | Remarks. | |
|---|---|---|---|---|---|---|---|---|---|---|---|
| tm. | Steam and Sail. | Total Stm. | Sail. | Grand Total Run. | | Tons. | Avge per Run. | Direction. | Force. | | |
| Mls. | Mls. | Mls. | Mls. | Mls. | Mls. | | Cwt. | | | | |
| 221 | 143 | 364 | 1,634 | 1,998 | 6·01 | ... | ... | | | | |
| ... | ... | ... | 151 | 151 | 6·29 | ... | ... | E. by N. to S. | 2.4 | Cir. str. cir. cu | Fine with rain squalls. |
| ... | ... | ... | 171 | 171 | 7·12 | ... | ... | E.—N.N.E.—E. | 3.5 | Cir str. cir cu. - | „ „ „ |
| ... | ... | ... | 123 | 123 | 5·12 | ... | ... | E. to N.—N.E. | 4.2.0.2 | Cumulus | Fine with heavy rain sqls. |
| ... | ... | ... | 112 | 112 | 4·70 | ... | ... | E. by N.—S.E. | 2.1.3 | Str. cu. cir. cu. | Frequent rain squalls. |
| ... | ... | ... | 146 | 146 | 6·06 | ... | ... | E.—to S.E. and E. | 2.3.4 | Cir. and cir. str. | Clear and steady light breeze. |
| ... | ... | ... | 152 | 152 | 6·33 | ... | ... | S.E. by S. to E.N.E. | 3.4 | Cir. cu. | Squally, with rain, to clear. |
| ... | ... | ... | 160 | 160 | 6·66 | ... | ... | N.E. to N. | 3.0.4 | „ „ | Rain squalls to clear. |
| ... | ... | ... | 164 | 164 | 6·83 | ... | ... | N.E. to N. | 4.5 | Cir. Ss. cir. cu. | Fine to wind and rain sqls. |
| ... | ... | ... | 181 | 181 | 5·46 | ... | ... | N.E.N.—N.E. | 3.4 | Stratus | Very fine; smooth water. |
| ... | ... | ... | 67 | 67 | 2·36 | ... | ... | N.E.N. by E.N.N.W. | 2.1.0.1 | „ | Light air with rain squalls. |
| ... | 81 | 81 | 123 | 123 | 5·12 | ... | ... | E.—N.E. | 2.4 | C. cir. strs. | Hvy. crs. swell, N.E.S. fine. |
| ... | 152 | 152 | 152 | 152 | 6·33 | ... | ... | N.E.—E.N.E. | 3.4 | Cir. cu. | Swell diminishing. |
| ... | ... | ... | 185 | 185 | 7·8 | ... | ... | N.E.—E.—E.N.E. | 4 | Cir. cu. cl. str. | Steady fresh brze., rain sqls. |
| ... | 13 | 13 | 162 | 175 | 6·48 | ... | ... | N.E. by N.—E.N.E. | 4 | Cir. cu. | Moderate breeze, fine. |
| 221 | 156 | 377 | 3,633 | 4,010 | 5·976 | | | | | | Per diem. 143·38 miles. |
| MARQUESAS GROUP. | | | | | | | | | | | |
| ... | 33 | 33 | ... | 33 | 5·5 | ... | ... | E.N.E. | 2.0 | Cir. cu. cir. str. | Fine, clear. |
| ... | 22 | 22 | ... | 22 | 5·87 | ... | ... | ... | ... | ... | ... |
| ... | ... | ... | ... | ... | ... | ... | ... | ... | ... | ... | Heavy willi-waws in & entering harbour. |
| ... | 75 | 75 | ... | 75 | 5 | ... | ... | E.—N.E. | 2.4 | Cir. str. stratus. | Fine. |
| 10 | ... | 10 | ... | 10 | 6·66 | ... | ... | W.N.W.—N.E. | 2.4 | Cirro. cu. | Fine. |
| 40 | ... | 40 | ... | 40 | 7 | ... | ... | E.—E.S.E. | 2.4 | Cirro. cu. | Willi-waws under Roa-Pous hills and rain sqls. |
| 50 | 130 | 180 | ... | 180 | 5·58 | ... | ... | ... | ... | ... | Pr. diem. 133·92.* |
| * This average includes lying-to off Cape Martin, Mouka-hiva, 6 hrs. for daylight. | | | | | | | | | | | |
| ... | ... | ... | ... | ... | ... | ... | ... | ... | ... | ... | ... |
| ... | ... | ... | 106 | 106 | 5·79 | ... | ... | E.—E.S.E. | 2.4 | Cirro. cu. | Fine with rain squalls. |
| ... | ... | ... | 149 | 149 | 6·20 | ... | ... | S.E. by W.—E.N.E. | 2.4 | Cirro. cu. cu. | Fine, clear with rain sqls. |
| ... | ... | ... | 170 | 170 | 7·08 | ... | ... | S.E. by S. | 4 | Cir. cu. cir. str. | Fine clear weather. |
| 38 | ... | 38 | 12 | 50 | ... | ... | ... | S.E. by E.—E. | 4.2 | Cum. nim. | Fine, rain sqls. |
| 219 | ... | 219 | ... | 219 | 9·12 | ... | ... | S. calm. | 1.2.0 | ... | Passed btwn. Rangiroa and Arutna Atolls, and in sight of them. |
| 30 | ... | 30 | ... | 30 | 6·66 | ... | ... | ... | ... | ... | ... |
| 287 | ... | 287 | 437 | 724 | 6·09 | 46·8 | 1·13 | 12 hours lying off & on for daylight at Ahii; & 5 hours' delay in approaching and landing | | | |
| SOCIETY ISLANDS | | | | | | | | | | | |
| 22 | ... | 22 | ... | 22 | 8 | ... | ... | N.E.—S.S.W. | 2 | Cirro. cu. | Fine, clear weather. |
| ... | 122 | 122 | ... | 122 | 7 | ... | ... | S.S.W.—S.E. by S. | 2.4 | Cir. cu. cir. str. | Fine, clear weather. |
| 30 | ... | 30 | ... | 30 | 6 | ... | ... | S E. | 0.2 | Cumulus. | Fine, clear weather. |
| 52 | 122 | 174 | ... | 174 | 6·89 | Time was lost making and entering the narrow passages through the coral reefs which encircle all these islands. | | | | | |

*Abstract from Log-book of "Wanderer," R.Y*

| 1880. | | | Temperature of — Watr. Deg. Fahr. | Temperature of — Air. Deg. Fahr. Ht. | Lt. | Position at Noon. Latitude. Deg. | Min. | Longitude. Deg. | Min. | Course. |
|---|---|---|---|---|---|---|---|---|---|---|
| Aug. 5 | Leave Cowes ... ... ... ... | 7·15 P.M. | 53 | 62 | 70 | ... | ... | ... | ... | Various |
| " 6 | High hd. sea; hosd. tpmsts. & Tt. Vd. | ... | 55 | 74 | 64 | 49 | 15 N. | 4 | 47 W. | W.S.W. |
| " 7 | High beam sea; ship rolling heavily | ... | 64 | 65 | 63 | 46 | 58 | 7 | 7 | S.W. |
| " 8 | Weather fining; sea going down; sight Cape P. ... ... ... ... | ... | 65 | 67 | 64 | 43 | 50 | 8 | 37 | S.W. |
| " 9 | Arrive at Vigo ... ... ... ... | 6·30 A.M. | 67 | 82 | 58 | ... | ... | ... | ... | Various |
| | *Duration of voyage, 3½ days.* | | | | | | | | | |
| | TOTALS ON VOYAGE ... ... ... | ... | ... | ... | | ... | ... | ... | ... | ... |
| " 11 | Leave Vigo ... ... ... ... | 11.35 A.M. | 62 | ... | | ... | ... | ... | ... | Various |
| " 12 | Arrive at Lisbon... ... ... ... | 11.35 A.M. | 63·44 | 71 | 66 | ... | ... | ... | ... | .. |
| | *Duration of voyage, 1 day.* | | | | | | | | | |
| " 21 | Leave Lisbon ... ... ... ... | 12 noon | 63·67 | ... | | ... | ... | ... | ... | Various |
| " 22 | ... ... | ... | 69·70 | 73 | 69 | 36 | 58 N | 11 | 12 W. | W.S.W |
| " 23 | ... ... | ... | 74·70 | 78 | 70 | 34 | 42 | 13 | 45 | ... |
| " 24 | ... ... | ... | ... | 78 | 69 | 33 | 29 | 15 | 20 | W. by S. ½ S. |
| " 25 | Arrive at Funchal ... ... ... | 12 noon | 75 | 76 | 69 | ... | ... | ... | ... | Various |
| | *Duration of voyage, 4 days.* | | | | | | | | | |
| | TOTALS ON VOYAGE ... ... ... | ... | ... | ... | | ... | ... | ... | ... | ... |
| " 27 | Leave Funchal ... ... ... ... | 11 A.M. | 79 | 79 | 74 | 32 | 33 N | 16 | 31 W. | S. by E. |
| " 28 | Feathered screw. Peak of Teneriffe in sight ... ... ... ... | ... | ... | 80 | 73 | 29 | 48 | 16 | 22 | S. by W. |
| " 29 | ... ... | ... | ... | 78 | 73 | 27 | 33 | 16 | 51 | Various |
| " 30 | ... ... | ... | ... | 75 | 83 | 25 | 16 | 18 | 22 | S. 31° 6' W. |
| " 31 | ... ... | ... | ... | 86 | 74 | 23 | 7 | 20 | 7 | S. 36° 29' W. |
| Sept. 1 | ... ... | ... | ... | 85 | 76 | 20 | 0 | 22 | 7 | S.W. by W. ½ W. |
| " 2 | ... ... | ... | ... | 85 | 75 | 18 | 28 | 24 | 36 | S. 45° W. |
| " 3 | ... ... | ... | 79 | 81 | 73 | 16 | 38 | 25 | 44 | Various |
| " 4 | Began to steam at ... ... ... | 9 A.M. | 80 | 86 | 79 | 14 | 35 | 25 | 46 | S. |
| " 5 | ... ... | ... | 80·81 | 79 | 81 | 12 | 47 | 23 | 33 | S. 46° E. |
| " 6 | ... ... | ... | 81 | 84 | 79 | 10 | 48 | 21 | 25 | 45 |
| " 7 | ... ... | ... | 80·79 | 85 | 79 | 9 | 30 | 18 | 52 | 63 |
| " 8 | ... ... | ... | 70·78 | 80 | 77 | 8 | 2 | 15 | 41 | 65 W |
| " 9 | ... ... | ... | 79·80 | 79 | 77 | 6 | 47 | 12 | 54 | 66 |
| " 10 | ... ... | ... | 79·80 | 79 | 75 | 5 | 41 | 10 | 17 | 70 |
| " 11 | ... ... | ... | 79 80 | 79 | 76 | 3 | 40 | 7 | 23 | 51 |
| " 12 | ... ... | ... | 79 | 79 | 74 | 3 | 12 | 4 | 12 | 81 |
| " 13 | ... ... | ... | 74·75 | 78 | 75 | 2 | 34 | 1 | 11 | 78 |
| " 14 | ... ... | ... | 75·77 | 80 | 77 | 2 | 21 | 1 | 23 E. | 76 |
| " 15 | ... ... | ... | 76 77 | 81 | 75 | 1 | 32 | 3 | 36 | 70 |
| " 16 | ... ... | ... | 77 | 80 | 77 | 0 | 45 | 6 | 29 | 76 |
| " 17 | Arrive at Gaboon ... ... ... | 2 P.M. | 79 | 79 | 76 | ... | ... | .. | ... | Various |
| | *Duration of voyage, 21½ days.* | | | | | | | | | |
| | TOTALS ON VOYAGE ... ... ... | ... | ... | ... | | ... | ... | ... | ... | ... |
| " 26 | Leave Gaboon ... ... ... ... | 6.30 A.M. | 79 | 82 | 81 | 0 | 8 N | 9 | 3 E. | Various |
| " 27 | ... ... | ... | 79 | 82 | 77 | 1 | 17 S. | 6 | 23 | S. 62° W. |
| " 28 | ... .. | ... | 76·79 | 81 | 76 | 3 | 27 | 4 | 52 | Various |
| " 29 | ... ... | ... | 76·74 | 79 | 74 | 6 | 15 | 3 | 48 | S. 21° 55' W. |
| " 30 | ... ... | ... | 78·71 | 76 | 72 | 9 | 1 | 2 | 35 | 22 |
| Oct. 1 | ... ... | ... | 72 | 71 | 69 | 11 | 0 | 0 | 15 | 49 30' |
| " 2 | ... ... | ... | 70 | 73 | 67 | 13 | 9 | 1 | 58 W. | 45 13 |
| " 3 | ... ... | ... | 70 | 67 | 65 | 15 | 16 | 4 | 56 | 46 30 |
| " 4 | Arrive at St. Helena ... ... ... | 6.30 A.M. | 70 | 68 | 66 | ... | ... | ... | ... | Various |
| | *Duration of voyage, 8 days.* | | | | | | | | | |
| | TOTALS ON VOYAGE ... ... ... | ... | ... | ... | | ... | ... | ... | ... | ... |
| " 7 | Leave St. Helena... ... ... ... | 10.35 A.M. | 68 | 64 | 78 | ... | ... | ... | ... | S. 83° W. |
| " 8 | ... ... | ... | ... | 65 | 73 | 15 | 48 S. | 7 | 27 W | — |
| " 9 | ... ... | ... | ... | 66 | 71 | 15 | 44 | 9 | 37 | N. 88° W. |
| " 10 | ... ... | ... | ... | 66 | 72 | 15 | 37 | 11 | 18 | 85° 57' |
| " 11 | ... ... | ... | ... | 67 | 74 | 15 | 24 | 13 | 12 | 83 |
| " 12 | ... ... | ... | 73 | 68 | 72 | 15 | 11 | 15 | 16 | 84 1 |
| " 13 | ... ... | ... | 73·74 | 71 | 74 | 14 | 57 | 17 | 57 | 84 41 |
| " 14 | ... ... | ... | 74·75 | 70 | 76 | 14 | 40 | 21 | 25 | 85 19 |
| " 15 | ... ... | ... | 75·74 | 71 | 78 | 14 | 23 | 23 | 46 | 83 |
| " 16 | ... ... | ... | 75 | 72 | 77 | 14 | 12 | 26 | 30 | 86 |
| " 17 | ... ... | ... | 77 | 74 | 79 | 13 | 59 | 29 | 24 | 85 |
| " 18 | ... ... | ... | 77 | 80 | 76 | 13 | 43 | 32 | 4 | 84 |
| " 19 | ... ... | ... | 78 | 81 | 75 | 13 | 19 | 34 | 50 | 81 |
| " 20 | Arrive at Bahia ... ... ... ... | 4 P.M. | 78 | 82 | 79 | ... | ... | ... | ... | Various |
| | *Duration of voyage, 13 days 6 hours.* | | | | | | | | | |
| | TOTALS ON VOYAGE ... ... ... | ... | ... | ... | | ... | ... | ... | ... | ... |

*ken from Place of Departure to Place of Arrival.*

| Distances run under Steam and Sail. | | | | | Average Speed per hour. | Coal used, with average per mile run under steam. | | Wind and Weather. | | Barometer. | | Remarks. |
|---|---|---|---|---|---|---|---|---|---|---|---|---|
| m. | Steam and Sail. | Total Stm. | Sail. | Grand Total Run. | | Tons. | Avge. per Run. | Direction. | Force. | Hst. | Lst. | |
| ls. | Mls. | Mls. | Mls. | Mls. | Mls. | | Cwts. | | | | | |
| ... | ... | ... | ... | ... | ... | ... | ... | ... | 0 | 29 96 | | |
| 160 | ... | 160 | ... | 160 | 10 | ... | ... | W.N.W. | 7 | 30·40 | 29·78 | |
| 226 | ... | 226 | ... | 226 | 9·41 | ... | ... | W. by S. | 7 | 29·87 | 29 90 | |
| 200 | ... | 200 | ... | 200 | 8 33 | ... | ... | W.S.W. | 6.3 | 29·34 | 30·22 | |
| 135 | ... | 135 | ... | 135 | 7·54 | 8¼ | ·913 | N. by E. | 8 0 | 30·50 | 30 12 | Stay in Vigo two days. |
| 721 | ... | 721 | ... | 721 | ... | ... | ... | ... | ... | 30·27 | 30 11 | |
| ... | ... | ... | ... | ... | ... | ... | ... | Calm. | ... | 30·09 | 30 02 | |
| 257 | ... | 257 | ... | 257 | 10·70 | 16·25 | 1·264 | ... | ... | 30 | 29·98 | Stay in Lisbon nine days. |
| ... | ... | ... | ... | ... | ... | ... | ... | W. by N. | 0.2 | 30·06 | | |
| 150 | ... | 150 | ... | 150 | 6·25 | ... | ... | S.W. to W. by N. | 2.5 | 30·04 | 30·19 | |
| ... | 177 | 177 | ... | 177 | 7·37 | ... | ... | W. by N. | 1.2.8 | 30·18 | 30·24 | |
| ... | ... | ... | 114 | 114 | 4·75 | ... | ... | W.N.W. | 3.4 | 30·18 | 30·23 | |
| 30 | ... | 30 | 66 | 96 | 4 | ... | ... | W.N.W.—N. by E. | 1.3.2 | 30·20 | 30·29 | |
| 180 | 177 | 357 | 180 | 537 | 5 59 | 11·50 | ·634 | ... | ... | 30 31 | 30·34 | |
| 11 | ... | 11 | ... | 11 | 10 | ... | ... | E N.E. | 2.1 | 30·30 | 30 27 | |
| ... | ... | ... | 172 | 172 | 7·11 | ... | ... | E N.E. | 4.3 2 | 30·27 | 30·21 | |
| ... | ... | ... | 150 | 150 | 6 25 | ... | ... | N.E. | 2.3 | 30·21 | 30·15 | |
| ... | ... | ... | 160 | 160 | 6·66 | ... | ... | E.N E | 2.3.4 | 30·10 | 30·19 | |
| ... | ... | ... | 160 | 160 | 6·66 | ... | ... | N.E. by E. E.N.E. | 4 3.2 | 30·13 | 30 13 | |
| ... | ... | ... | 180 | 180 | 7·5 | ... | ... | E N.E. N.E. by S. | 4 2.5 | 30·10 | 30·14 | |
| ... | ... | ... | 194 | 194 | 8·08 | ... | ... | N.E. by E —E. | 2 4 4.2 | 30·10 | 30·19 | |
| ... | ... | ... | 139 | 139 | 5·78 | ... | ... | E. to E N.E. | 2.3 | 30 6 | 30·12 | |
| ... | 23 | 23 | 99 | 122 | 5 | ... | ... | S.E.—N.E.—E. | 1.2 1 | 30·7 | 30·13 | |
| ... | 182 | 182 | ... | 182 | 7 6 | ... | ... | S.E.—E. by N. nim. | 1 3.5 | 30 7 | 30 11 | |
| 173 | ... | 173 | ... | 173 | 7·20 | ... | ... | N.W. to S.E. nim. | 1 2 0 | 30·4 | 30·10 | |
| ... | 173 | 173 | ... | 173 | 7 20 | ... | ... | S.W.—S.S.W. nim. | 2.5 | 30 | 30·05 | |
| ... | 208 | 208 | ... | 208 | 8·66 | ... | ... | S.W —W.S.W. nim. | 3.5 | 30 | 30·09 | |
| ... | 181 | 181 | ... | 181 | 7·54 | ... | ... | W S.W. nim. | 2.5 | 30·02 | 30·29 | |
| 160 | ... | 160 | ... | 160 | 6·66 | ... | ... | W.S.W. to S nim. | 5 | 30·05 | 30 15 | |
| ... | 194 | 194 | ... | 194 | 8·08 | ... | ... | S.W. by S. nim. | 5.6 | 30·14 | 30 06 | |
| ... | 16 | 16 | 177 | 193 | 8 08 | ... | .. | S.W. to S.S W. cirro. cu | 2.5 | 30 | 30·12 | |
| ... | ... | ... | 184 | 184 | 7·7 | ... | ... | S. by W. | 2.5 | 30 | 30·19 | |
| ... | ... | ... | 155 | 155 | 6·45 | ... | ... | S by W. to S.S.W. cu. st | 2 | 30·01 | 30 12 | |
| ... | ... | ... | 142 | 142 | 5 91 | ... | ... | S.S.W. cu. st & cirro. cu. | 2.4 | 30·21 | 30·05 | |
| ... | ... | ... | 200 | 200 | 8·33 | ... | ... | S.W. cirro. cu. & nim. | 4 | 30·14 | 30·04 | |
| 206 | ... | 206 | 10 | 216 | 8·30 | ... | ... | S.S.W. nim. | 5.6 | 30·02 | 30·19 | |
| 550 | 977 | 1,527 | 2,122 | 3,649 | ·7·20 | 52 25 | ·681 | | | | | |
| 30 | ... | 30 | ... | 30 | 5 | ... | ... | S.S.W. cu. stratus | 4 | 30·08 | ... | |
| 181 | ... | 181 | ... | 181 | 7·54 | ... | ... | W.S.W cu. stra. | 2.4 | 30·03 | 30·12 | |
| 169 | ... | 169 | ... | 169 | 7·04 | ... | ... | S.S.W.—W. by S. nim. | 4.5 | 30·03 | 30 12 | |
| 184 | ... | 184 | ... | 184 | 7·66 | ... | ... | S.W. to S. by W. cu. | 4.5 | 30·02 | 30·11 | |
| 179 | ... | 179 | ... | 179 | 7·87 | ... | ... | S.W.—S. | 2.4 | 30·03 | 30 09 | |
| 133 | .. | 133 | 49 | 182 | 7·58 | ... | ... | S.S W. cu. & nim. | 4 5 | 30·02 | 30·13 | |
| ... | ... | ... | 183 | 183 | 7 62 | ... | ... | S. to S.S E. | 5 | 30 05 | 30·19 | |
| ... | ... | ... | 214 | 214 | 8·91 | ... | ... | S.S E cirro & cu. | 5.7 | 30·07 | 30 21 | |
| 30 | ... | 30 | 30 | 60 | ... | ... | ... | S.S.E. to S.E. | 5 | 30·15 | 30 21 | Lying-to all night for daylt. |
| 906 | ... | 906 | 476 | 1,382 | 7 17 | 40·50 | ·894 | | | | | |
| ... | ... | ... | 8 | 8 | ... | ... | ... | S.S.E. to S.E. co. cu. | 1 2 4.1 | 30·12 | 30·15 | To clear the ships only. |
| ... | ... | ... | 97 | 97 | 4·04 | ... | ... | S.E. to S.E. by E. cum. | 2.4.2 | 30·10 | 30·15 | |
| ... | ... | .. | 131 | 131 | 5·45 | ... | ... | S.E. to E S.E ni. cu. | 2 | 30·10 | 30·15 | |
| .. | ... | ... | 99 | 99 | 4·11 | ... | ... | S. to S.S.E cumulus | 2 0 | 30·10 | 30 20 | |
| ... | ... | ... | 101 | 101 | 4·20 | ... | ... | S E. by E co. cu. | 2 3 | 30·10 | 30 17 | |
| ... | ... | ... | 124 | 124 | 5 16 | ... | ... | E. by S. S.S.W. E.S.E. cu | 4.2 | 30 12 | 30 19 | |
| ... | 118 | 118 | 27 | 145 | 6·04 | ... | ... | S S E to N W. ni. cu. | 4.3 | 30·14 | 30 21 | |
| ... | 209 | 209 | ... | 208 | 8·66 | ... | ... | S E by E.—E.S E co cu | 2.4.2 | 30·16 | 30·22 | |
| ... | 120 | 120 | 14 | 134 | 5 60 | ... | ... | E.S.E co. cu. & str. | 4.2.5 | 30 15 | 30·23 | |
| ... | ... | ... | 163 | 163 | 6·80 | ... | ... | E.S E co. str. | 4 | 30·10 | 30 19 | |
| | ... | ... | 163 | 162 | 6 80 | ... | ... | E.S.E. cumulus. | 4 | 30 10 | 30·17 | |
| ... | ... | ... | 161 | 161 | 6·70 | ... | ... | E by S.—N.E. co. cu. | 4.5 | 30·10 | 30 15 | |
| ... | ... | ... | 161 | 161 | 6 70 | ... | ... | N.E. by N cumulus. | 4 | 30 06 | 30·16 | |
| 210 | ... | 210 | 14 | 224 | 8 | ... | ... | Various. cu. cir. str. | 0 2 | 30 10 | 30·17 | |
| 210 | 446 | 656 | 1,263 | 1,919 | 6·08 | 28·25 | ·861 | E. by S.—E.S.E. | 4.2 | 30·09 | 30·15 | |

*Abstract from Log-book of "Wanderer," R.Y.*

| 1880. | | | Temperature of Watr. Deg. Fahr. | Air Deg. Fahr. Ht. | Air Deg. Fahr. Lt | Position at Noon. Latitude. Deg. | Latitude. Min | Longitude. Deg. | Longitude. Min | Course. |
|---|---|---|---|---|---|---|---|---|---|---|
| Oct. 26 | Leave Bahia ... ... ... ... | 5 P.M. | 80 | 80 | 75 | ... | ... | ... | ... | Various |
| ,, 27 | ... ... | ... | 78 | 80 | 73 | 15 | 16 S. | 38 | 11 W. | ... |
| ,, 28 | High sea; ship pitching heavily ... | ... | 76 | 77 | 75 | 17 | 10 | 38 | 14 | S. 1° E |
| ,, 29 | ,, ,, rolling ... ... ... | ... | 75 | 82 | 74 | 20 | 6 | 38 | 30 | 0 30′ W |
| ,, 30 | ... ... | ... | 72 | 83 | 73 | 22 | 44 | 40 | 59 | 42° |
| ,, 31 | Arrive at Rio Janeiro ... ... ... | 8 A.M. | 70 | 74 | 86 | ... | ... | ... | ... | Various |
| | Lying-to for daylight. | | | | | | | | | |
| | *Duration of voyage, 4 days 15 hours.* | | | | | | | | | |
| | TOTALS ON VOYAGE ... ... ... | ... | ... | ... | | ... | ... | ... | ... | — |
| Nov. 13 | Leave Rio... ... ... ... ... | 6 P.M. | ... | ... | | ... | ... | ... | ... | ... |
| ,, 14 | Calm and light winds ... ... .. | ... | 75 | 81 | 72 | 25 | 10 S. | 44 | 34 W. | S. 31° W |
| ,, 15 | Fine night and nice fair breeze; all sqr. sails set ... ... ... ... | ... | 73 | 80 | 74 | 28 | 10 | 46 | 51 | 34 |
| ,, 16 | Heavy head sea ... ... ... ... | ... | 73 | 75 | 69 | 30 | 26 | 48 | 48 | 36 |
| ,, 17 | ... ... | ... | 67 | 63 | 62 | 32 | 7 | 49 | 50 | 28 |
| ,, 18 | Strong breeze and high head sea ... | ... | 61 | 62 | 57 | 33 | 21 | 51 | 27 | 49 |
| ,, 19 | Light breeze; sea diminishing ... | ... | 64 | 62 | 55 | 35 | 18 | 54 | 46 | 58 |
| ,, 19 | Arrive at Monte Video ... ... ... | 8 P.M. | — | 73 | 59 | ... | ... | ... | ... | Various |
| | *Duration of voyage, 6 days.* | | | | | | | | | |
| | TOTALS ON VOYAGE ... ... ... | ... | ... | ... | | ... | ... | ... | ... | — |
| ,, 22 | Leave Monte Video ... ... ... | 3 P.M. | 65 | 76 | 68 | ... | ... | ... | ... | Various |
| ,, 23 | Arrive at Buenos Ayres ... ... | 7 A.M. | 69 | ... | | ... | ... | ... | ... | — |
| | *Duration of voyage, 15 hours.* | | | | | | | | | |
| ,, 30 | Leave Buenos Ayres ... ... ... | 5 P.M. | 67 | 71 | 60 | ... | ... | ... | ... | — |
| Dec. 1 | Arrive at Monte Video ... ... ... | 7 A.M. | 65 | 66 | 60 | ... | ... | ... | ... | — |
| | *Duration of voyage, 14 hours.* | | | | | | | | | |
| ,, 4 | Leave Monte Video ... ... ... | 5 P.M. | 68 | 77 | 62 | ... | ... | ... | ... | — |
| ,, 5 | Lovely day ... ... ... ... | ... | 68·61 | 76 | 67 | 37 | 31 S. | 56 | 30 W. | — |
| ,, 6 | Heavy N.E. swell ... ... ... | ... | 57 | 65 | 60 | 39 | 54 | 59 | 2 | S. 40 W |
| ,, 7 | Fine and clear ... ... ... ... | ... | 56 | 68 | 51 | 41 | 50 | 61 | 53 | 48 |
| ,, 8 | Arrive at Pyramid Bay, New Gulf ... | 10 A.M. | 56 | 72 | 61 | ... | ... | ... | ... | Various |
| ,, 9 | Run across Gulf to Port Madryn ... Sailed 6 A.M., arrived 10 P.M. | ... | 60 | 79 | 62 | ... | ... | ... | ... | — |
| ,, 10 | Heavy gale from N.E. on dead lee shore ... ... ... ... | ... | 59 | — | | ... | ... | ... | — | — |
| ,, 11 | Return and arrive at Port Madryn ... | 9 A.M. | 57·60 | 70 | 68 | ... | ... | ... | ... | — |
| | *Time under way, 14 hours.* | | | | | | | | | |
| ,, 11 | Leave Port Madryn again ... ... | 4 15 P.M. | 57·60 | ... | | ... | ... | ... | ... | ... |
| ,, 12 | Anchor off Chuput river ... ... | 6.15 A.M. | 57 | 64 | 59 | ... | ... | ... | ... | Various |
| ,, 13 | Leave Chuput ... ... ... ... | 5 A.M. | 56 | 57 | 61 | 44 | 27 | 64 | 50 | S. 5° E. |
| ,, 14 | ... ... ... ... | ... | 50 | 61 | 55 | 47 | 26 | 65 | 10 | 5 W. |
| ,, 15 | ... ... ... ... | ... | 46 | 59 | 53 | 51 | 01 | 66 | 53 | 18 |
| ,, 16 | Anchor under Condor Cliff ... ... | 6.30 A.M. | 46 | 54 | 44 | ... | ... | ... | ... | Various |
| | *Duration of voyage, 8 days.* | | | | | | | | | |
| | TOTALS ON VOYAGE ... ... ... | ... | ... | ... | | ... | ... | ... | ... | — |
| ,, 19 | Leave Cape Virgins ... ... ... | 2.30 A.M. | ... | ... | | ... | ... | ... | ... | ... |
| | Anchor at Sandy Point ... ... | 4.30 P.M. | 46 | 48 | 54 | ... | ... | ... | ... | Various |
| ,, 22 | Leave Sandy Point ... ... ... | 10.50 A.M. | 46 | 44 | 54 | ... | ... | ... | ... | — |
| | Anchor off Elizabeth Island ... | 2.?0 P.M. | | | | | | | | |
| ,, 27 | Leave Elizabeth Island ... ... | 6 A.M. | 46 | 56 | 44 | ... | ... | ... | ... | — |
| | Anchor at Sandy Point ... ... | 9 P.M. | | | | | | | | |
| ,, 30 | Leave Sandy Point ... ... ... | 4.10 A.M. | 47 | 61 | 47 | ... | ... | ... | ... | — |
| | Anchor at Borja Bay ... ... | 7.10 P.M. | | | | | | | | |
| ,, 31 | Leave Borja Bay... ... ... ... | 4.15 A.M. | 48 | 44 | 55 | ... | ... | ... | ... | ... |
| 1881. | Anchor Coal Mine, Skyring Water | 5.45 P.M. | | | | | | | | |
| Jan. 1 | Leave Skyring Water ... ... ... | 4 A.M. | 50 | 50 | 45 | ... | ... | ... | ... | — |
| | Havannah Point ... ... ... | 8 P.M. | | | | | | | | |
| ,, 2 | From off Havannah Point ... ... | 6 A.M. | 48 | 45 | 52 | ... | ... | ... | ... | — |
| | Anchor at Isthmus Bay ... ... | 3.30 P.M. | | | | | | | | |
| ,, 3 | Leave Isthmus Bay ... ... ... | 6.30 A.M. | 49 | 47 | 57 | ... | ... | ... | ... | ... |
| | Anchor at Puerto Bueno ... ... | 6 30 P.M. | | | | | | | | |
| ,, 5 | Leave Puerto Bueno ... ... ... | 4.30 A.M. | 49 | 57 | 50 | ... | ... | ... | ... | ... |
| | Anchor at Port Grappler ... ... | 7 P.M. | | | | | | | | |
| ,, 6 | Leave Port Grappler ... ... ... | 8.30 A.M. | 50 | 46 | 51 | ... | ... | ... | ... | ... |
| | Return to Port Grappler, going round Saumarez Island ... ... ... | 12 15 P.M. | | | | | | | | |
| ,, 7 | Leave Port Grappler ... ... ... | 5 A.M. | 49 51 | 51 | 40 | ... | ... | ... | ... | — |
| | Sombrero Islands abeam ... ... | 8 P.M. | | | | | | | | |
| | *Time under way, 5 days 4½ hours.* | | | | | | | | | |
| | TOTALS IN STRAITS ... ... ... | ... | ... | ... | | ... | ... | ... | ... | ... |

*ı from Place of Departure to Place of Arrival.*

| ...ances run under Steam and Sail | | | | Average Speed per hour. | Coal used, with average per mile run under steam. | | Wind and Weather. | | Barometer | | Remarks. |
|---|---|---|---|---|---|---|---|---|---|---|---|
| Steam and Sail. | Total Stm. | Sail. | Grand Total run. | | Tons. | Avge per Run. | Direction. | Force | Hst. | Lst. | |
| Mls. | Mls. | Mls. | Mls. | Mls. | | Cwts | | | | | |
| ... | ... | ... | ... | ... | ... | ... | E. by S.—S.S.E. cum. | 4·2 | 30·17 | 30·24 | |
| 135 | 135 | ... | 135 | 7·01 | ... | ... | S.E. by S.—S.S.E. nim. | 2·5 | 30 21 | 30 25 | |
| 120 | 120 | ... | 120 | ·5 | ... | ... | S.E. nim | 5·7 | 30·23 | 30 32 | |
| 176 | 176 | ... | 176 | 7 88 | ... | ... | S.E.—E cum. str. | 6·7 | 30·25 | 30 31 | |
| 209 | 209 | ... | 209 | 8 70 | ... | ... | E.N.E.—N.E. cirro. cu. | 4 6 | 30 24 | 30·18 | |
| ... | 113 | ... | 113 | 5 65 | ... | ... | N.E. | 0·2 | 29·99 | 29·91 | Very heavy thunderstorm and floods of rain. |
| 640 | 753 | ... | 753 | 6·78 | 28·50 | ·624 | | | | | |
| ... | ... | ... | ... | ... | ... | ... | ... | ... | 30 | ... | A glorious evening & night. |
| ... | 156 | ... | 156 | 8·66 | ... | ... | S.W. by S. cirro. stra. | 0·2 | 30 | 30·04 | |
| 218 | 218 | ... | 218 | 9·08 | ... | ... | S. by E. to N. by E. nim. | 2·4 | 30·02 | 29 91 | |
| 170 | 170 | ... | 170 | 7·08 | ... | ... | N. to S.W. by S. nim. | 6·8 | 29 76 | 30·06 | |
| 115 | 115 | ... | 115 | 4·80 | ... | ... | S. by W. to S.W cu. st. cl | 4·6 | 30 15 | 30·25 | |
| ... | 113 | ... | 113 | 4·70 | ... | ... | S.W. by S. cirrus. | 4·6 | 30 22 | 30·27 | |
| 119 | 119 | ... | 119 | 8·30 | ... | ... | S.W. to N.W. cirro. cu. | 2 5 | 30·20 | 30 12 | |
| ... | 74 | ... | 74 | 9·25 | ... | ... | N. | 2·0 | 30 05 | 29·96 | |
| 702 | 1,045 | ... | 1,045 | 7·15 | 41 | ·669 | | | | | |
| ... | ... | ... | ... | ... | ... | ... | N W.—N.N.W. cirr. | 4 | 30·15 | 30·10 | |
| ... | 115 | ... | 115 | 7·66 | 7·75 | 1·348 | N.W.—N.N.W. cum. | 4·6 | 30·07 | 30 04 | |
| ... | ... | ... | ... | ... | ... | ... | E. by N. to N. cirro. cu. | 2·7 | 30·28 | 30 25 | |
| ... | 115 | ... | 115 | 8·21 | 7·25 | 1·260 | N. cirro. str. | 4·7 | 30 25 | 30·26 | |
| ... | ... | ... | ... | ... | ... | ... | W. cumulus | 2 0 | 29·82 | 29·87 | |
| ... | 160 | ... | 160 | 8·42 | ... | ... | S.W.—S.S E. cir. str. | 0 2 | 29·80 | 29 99 | |
| 186 | 186 | ... | 186 | 7·75 | ... | ... | W. cir. str. | 2·5 | 29·97 | 29·80 | |
| 173 | 173 | ... | 173 | 7·20 | ... | ... | S.W.—N E cir. cu. | 2·5 | 29·86 | 29 60 | |
| 133 | 133 | ... | 133 | 6 | ... | ... | W.—W N.W. cir. cu. | 0·7 | 29·49 | 29·84 | Slowed down for daylight. |
| ... | 85 | ... | 85 | 8·75 | ... | ... | E.N.E. cir. str. | 4·5 | 29·82 | 30·20 | |
| ... | ... | ... | ... | ... | ... | ... | ... | ... | 30·10 | 29·85 | Got under way 7 P.M., wind veering round compass. Steamed dead slow across gulf & back with S.W. wd. |
| ... | 60 | ... | 60 | 4·30 | ... | ... | Round compass | 7·8 | 29·59 | 29·69 | |
| ... | ... | ... | ... | ... | ... | ... | N.E. cirro. cu. | 5 | 29·65 | 30·0 | |
| ... | 75 | ... | 75 | 5·35 | ... | ... | S.W. to E. cirro. cu. | 5·7 | 29 94 | 29 64 | |
| 59 | 59 | ... | 59 | 8·43 | ... | ... | W.S.W. cumulus | 2·6 | 29·70 | 29·90 | |
| 180 | 180 | ... | 180 | 7·50 | ... | ... | S.—N.W. cir. str. | 6·2 | 29·94 | 29·89 | |
| 225 | 225 | ... | 225 | 9 37 | ... | ... | N.W.—W.S.W. cum. | 4·6 | 29 70 | 29·08 | |
| ... | 108 | ... | 108 | 5·79 | ... | ... | W.S.W cum. | 8·9 | 28 95 | 28·84 | Furious gale from W.S.W. to S.W.; ride it out 16th, 17th, 18th. Lowest glass, 28·65. |
| 956 | 1,394 | ... | 1,394 | 7·18 | 58 25 | ·764 | Furious gale, S.W. | 8·9 | 28 98 | 29·04 | |
| ... | ... | ... | ... | ... | ... | ... | cirro. cu. | 5·5 | ... | ... | Stiff S.W. breeze. |
| ... | 124 | ... | 124 | 7·75 | ... | ... | S.W. cirro. cum. | 4 6 | 29·35 | 29·47 | |
| ... | 24 | ... | 24 | 6 | ... | ... | W. cumulus | 2·5 | 29·74 | 29 96 | |
| ... | 24 | ... | 24 | 8 | ... | ... | N.W.—N.E. cu. cir. cu. | 5 | 29·62 | 29 45 | |
| ... | 104½ | ... | 104½ | 6·96 | ... | ... | N. to W. | 2 | 29·61 | 29·34 | |
| ... | 91 | ... | 91 | 6 74 | ... | ... | W. to N.W. cu. nim. | 2·5 | 29·32 | 29·42 | Aiding *Santos*. |
| ... | 127½ | ... | 127½ | 8 | ... | ... | W.—N.W. nim. | 0·4 | 29·42 | 29·52 | Stmg. slow off & on all night. |
| ... | 85½ | ... | 85½ | 9 | ... | ... | N.W. nim. cu. | 5·2 | 29·48 | 29 91 | Very thick and raining hard. |
| ... | 82 | ... | 82 | 6·83 | ... | ... | N.W. cumulus | 5 | 29·94 | 29·72 | |
| ... | 111½ | ... | 111½ | 7 67 | ... | ... | ... | 6·2 | 29·42 | 29·71 | |
| ... | 46 | ... | 46 | 8 | ... | ... | nim. | 2·7 | 29·71 | 29·48 | Down Chasm Reach, returning by Tay Reach. |
| ... | 97½ | ... | 97½ | 6·50 | ... | ... | ... | 5·7 | 29·37 | 29·74 | |
| ... | 917 | ... | 917 | 7·38 | | | | | | | |

*Abstract from Log-book of "Wanderer,"* i.

| 1881. | | | Temperature of Watr. Deg. Fahr. | Temperature of Air. Deg. Fahr. (Ht. Lt.) | Position at Noon. Latitude. Deg. | Latitude. Min. | Longitude. Deg. | Longitude. Min. | Course. |
|---|---|---|---|---|---|---|---|---|---|
| Jan. 8 | From Sombrero yesterday at ... | 8 P.M. | 52·57 | 49 59 | 46 | 41 S. | 76 | 23 W | Various |
| ,, 9 | ... ... ... | ... | 55·60 | 55 64 | 43 | 23 | 75 | 51 | N. 7° E |
| ,, 10 | ... ... ... | ... | 60 62 | 62 59 | 39 | 54 | 74 | 38 | N. 15 E. |
| ,, 11 | ... ... ... | ... | 62 | 60 68 | 36 | 22 | 74 | 03 | N. 7 E |
| ,, 12 | Made fast to buoy Valparaiso ... | 1 P.M. | 56 | 59 72 | ... | ... | ... | ... | Various |
| | Left Valparaiso ... ... ... | 6 P.M. | 59·61 | | | | | | |
| ,, 13 | Anchor in Coquimbo ... ... | 3 P.M. | 59·61 | 61 70 | ... | ... | ... | ... | Various |

*Duration of voyage, 5 days 14 hrs.*

TOTALS ON VOYAGE ... ... ... | ... | ... | ... | ... | ... | ... | ... | —

WORK DONE I…

| 1881. | | | Watr. | Air | Lat. Deg. | Lat. Min. | Long. Deg. | Long. Min. | Course. |
|---|---|---|---|---|---|---|---|---|---|
| Feb. 15 | Left Coquimbo ... ... ... | Noon | 64 | | | | | | |
| ,, 16 | Arrived at Valparaiso ... ... | 4 45 P.M. | 60 63 | ... | ... | ... | ... | ... | Southward |
| Mar. 2 | Left ,, ... ... ... | 10.30 A.M. | 58·60 | ... | ... | ... | ... | ... | Northerly |
| ,, 3 | Arrived at Tongoy ... ... ... | 9 A.M. | 59·62 | ... | ... | ... | ... | ... | — |
| ,, 3 | Left ,, ... ... ... | 9 30 P.M. | | | | | | | |
| ,, 3 | Arrived at Coquimbo ... ... | 3.40 P.M. | ... | ... | ... | ... | ... | ... | Various |
| ,, 26 | Left ,, ... ... ... | Noon | 62·58 | | | | | | |
| ,, 27 | Arrived at Valparaiso ... ... | 8 30 P.M. | 60·59 | ... | ... | ... | ... | ... | — |
| April 3 | Left ,, ... ... ... | 6.30 A.M. | 60 61 | | | | | | |
| ,, 4 | Returned to ,, ... ... ... | 8.30 A.M. | 60 | | | | | | |
| ,, 11 | Left ,, ... ... ... | 3.45 P.M. | 61 | ... | ... | ... | ... | ... | — |
| ,, 12 | Returned to ,, ... ... ... | 10.15 A.M. | 60,61,62 | ... | ... | ... | ... | ... | — |
| ,, 13 | Left ,, ... ... ... | 4 P.M. | 62 | 61 62 | ... | ... | ... | ... | Various |
| ,, 14 | Arrived at Coquimbo ... ... | Noon | 62 | 65 62 | ... | ... | ... | ... | — |
| May 14 | Left ,, ... ... .. | 10 A.M. | 57,58,59 | 61 70 | ... | ... | ... | ... | Various |
| ,, 15 | Arrived at Valparaiso ... ... | 10 A.M. | 59 | 58 63 | | | | | |
| ,, 16 | Left ,, ... ... ... | Noon | 59 | 58 66 | ... | ... | ... | ... | Various |
| ,, 17 | At sea ... ... ... ... ... | ... | 59 | 61 60 | 32 | 58 S | 74 | 38 W. | N. 89 W |
| ,, 18 | ,, ... ... ... ... | ... | 60·61 | 57 60 | 32 | 50 | 76 | 41 | N. 86 W |
| ,, 19 | Arrived at Cumberland Bay, Juan Fernandez ... ... ... ... | 4 P.M. | 60,61,62 | 61 63 | ... | ... | ... | ... | S. 70 W. |
| ,, 20 | Left ,, ,, ... ... | Noon | 61 | 59 61 | ... | ... | ... | ... | — |
| ,, 21 | At sea ... ... ... ... ... | ... | 60·61 | 63 57 | 33 | 19 | 77 | 40 | Var. N. [illegible] E |
| ,, 22 | ,, ... ... ... ... | ... | 61 59 | 58 63 | 33 | 12 | 75 | 34 | N. 87 E |
| ,, 23 | ,, ... ... ... ... | ... | 59 | 62 64 | 33 | 4 | 73 | 20 | N. 75 E |
| ,, 24 | Arrived at Valparaiso ... ... | 1 P.M. | 59·58 | 61 64 | ... | ... | ... | ... | Various |
| ,, 27 | Left ,, ... ... ... | 9.20 P.M. | | | | | | | |
| ,, 29 | Returned to ,, ... ... ... | 2.10 P.M. | 59·63 | 57 | ... | ... | ... | ... | Various |
| ,, 31 | Left ,, ... ... ... | 4 P.M. | 57·59 | 56 58 | ... | ... | ... | ... | — |
| June 1 | Arrived at Coquimbo ... ... | 1 P.M. | 60·63 | 58 57 | ... | ... | ... | ... | Various |
| ,, 7 | Left ,, ... ... ... | 7 P.M. | 55·57 | 58 | | | | | |
| ,, 8 | Arrived at Valparaiso ... .. | 8.30 P.M. | 59·62 | 58,59,60 | ... | ... | ... | ... | Various |
| ,, 8 | Left ,, ... ... ... | 4.30 P.M. | 61·65 | ... | | | | | |
| ,, 9 | Arrived at Coquimbo ... ... | 2.30 P.M. | 62 67 | 59,60,57 | ... | ... | ... | ... | Various |

*Time under way, 19 days 22 hours.*

TOTALS OF NAVIGATION IN CHILI ... | ... | ... | ... | ... | ... | ... | ... | —

VOYAGE CONTIN…

| 1881. | | | Watr. | Air | Lat. Deg. | Lat. Min. | Long. Deg. | Long. Min. | Course. |
|---|---|---|---|---|---|---|---|---|---|
| June 16 | Leave Coquimbo ... ... ... | 4 P.M. | 60·59 | 56 59 | ... | ... | ... | ... | — |
| ,, 17 | At sea ... ... ... ... ... | Noon | 59·60 | 59 | 30 | 37 S | 73 | 16 W | N. [illegible] W |
| ,, 18 | ,, ... ... ... ... | ... | 58·56 | 61 | 30 | 37 | 77 | 10 | West |
| ,, 19 | ,, ... ... ... ... | ... | 60·57 | 61 | 30 | 11 | 80 | 59 | N. 83 W. |
| ,, 20 | ,, ... ... ... ... | ... | 64·58 | 63 | 29 | 49 | 84 | 27 | N. 83 W |
| ,, 21 | ,, ... ... ... ... | ... | 60·58 | 63 | 29 | 43 | 86 | 40 | N. 87 W |
| ,, 22 | ,, ... ... ... ... | ... | 63·58 | 63 64 | 29 | 27 | 88 | 19 | N. 80 W |
| ,, 23 | Very high beam sea, southerly ... | ... | 63·57 | 66 | 29 | 8 | 91 | 29 | N. 84 W |
| ,, 24 | ,, ,, ,, ... | ... | 65·60 | 66 | 27 | 59 | 93 | 35 | N. 54 W |
| ,, 25 | ,, ,, ,, ... | ... | 61·71 | 66 | 26 | 13 | 94 | 33 | N. 24 W. |
| ,, 26 | ,, ,, ,, ... | ... | 65·71 | 68 | 25 | 0 | 96 | 18 | N. 52 W |
| ,, 27 | Swell moderated ... ... ... | ... | 65·71 | 68 | 24 | 7 | 97 | 52 | N. 55 W |
| ,, 28 | Swell greatly reduced ... ... | ... | 65·76 | 70 | 23 | 8 | 99 | 58 | N 62 W |
| ,, 29 | Swell quite lost ... ... ... | ... | 67·78 | 71 | 22 | 16 | 102 | 40 | N. 71 W. |
| ,, 30 | ,, ... ... ... ... | ... | 67·78 | 71 | 21 | 22 | 105 | 33 | N. 71 W |

CARRIED FORWARD ... ... | ... | ... | ... | ... | ... | ... | ... | —

*n from Place of Departure to Place of Arrival.*

| stances run under Steam and Sail. | | | | Average Speed per hour. | Coal used, with average per mile run under steam. | | Wind and Weather. | | Remarks. | | |
|---|---|---|---|---|---|---|---|---|---|---|---|
| | | | | | | | | | Barometer | | |
| Steam and Sail. | Total Stm. | Sail. | Grand Total Run. | | Tons. | Avge. per Run | Direction. | Force | Hst. | Lst. | |
| Mls. | Mls | Mls. | Mls | Mls. | | Cwts | | | | | |
| ... | 66 | ... | 66 | 4 12 | ... | ... | S.W.—N.W. cir. cu. | 5 | 29·79 | 29·83 | Very high sea. |
| 200 | 200 | ... | 200 | 8·33 | ... | ... | N.E. cumulus | 4·2 | 29 64 | 29·77 | Long swell from S.E. |
| 216 | 216 | ... | 216 | 9 | ... | ... | W.N.W. cir. str. nim. | 2·4 | 29·82 | 30·08 | |
| 213 | 213 | ... | 213 | 8·87 | ... | ... | N.W.—W.S.W. cu. | 2·6 | 30·11 | 30·24 | |
| ... | 222 | ... | 222 | 8·88 | ... | ... | S. cir. str. | 2·6 | 30·21 | 30·01 | |
| ... | 180 | ... | 180 | 8·77 | ... | ... | Calm cir. cu. | 0 | 30·03 | 30·17 | Long swell S. |
| 629 | 1,097 | ... | 1,097 | 8·18 | 113·50 | 1 127 | | | | | |

R STAY IN CHILI.

| Steam and Sail. | Total Stm. | Sail. | Grand Total Run. | Average Speed per hour. | Tons. | Avge. per Run | Direction. | Force | Hst. | Lst. | Remarks. |
|---|---|---|---|---|---|---|---|---|---|---|---|
| ... | 200 | ... | 200 | 6·89 | ... | ... | S. | 2·4 | ... | | Heavy head sea. |
| 180 | 180 | ... | 180 | 8 | ... | ... | S.W. | 5 | ... | | Very fine. |
| ... | 21 | ... | 21 | 7 | ... | ... | Calm | 0 | ... | | Very fine. |
| ... | 200 | ... | 200 | 5 33 | ... | ... | S.W. | 5·6 | ... | | Fires banked while in Port. High head sea. |
| 50 | 50 | ... | 50 | 2 | ... | ... | N.W. gale | 7 | ... | | Steam and try-sails. |
| ... | ... | ... | ... | ... | ... | ... | N.W. gale | 6·7 | ... | | Steam and try-sails. |
| 40 | 40 | ... | 40 | 2 15 | ... | ... | N.W. gale | ... | ... | | Dirty and heavy rain. |
| ... | ... | ... | ... | ... | ... | ... | ... | ... | ... | | Fine and clear. |
| 200 | 200 | ... | 200 | 10 | ... | ... | S.W.—S.E.—N.E. | 2 | ... | | Lovely weather. |
| ... | 200 | ... | 200 | 8·33 | ... | ... | S. to W. | 3 | ... | | Fine. |
| ... | 16 | 135 | 151 | 6·29 | ... | ... | S.S.E.—S.—S.W. | 4·2 | ... | | Moderate, fine, squalls. |
| ... | ... | 96 | 96 | 4 | ... | ... | S. and variable. | 2 | ... | | Ditto, and strong N. currt. |
| 113 | 113 | 6 | 119 | 4·25 | ... | ... | S.W. to S.E. | 2 | ... | | Ditto, with showers. |
| ... | ... | ... | ... | ... | ... | ... | ... | ... | ... | | Two hrs. lying off and on in English Bay, while party vistd. Alex. Selkirk's cave. |
| 12 | 12 | 56 | 68 | 3 | ... | ... | S. to calm. | 0 3 | ... | | Fine, cloudy. |
| 9 | 91 | 12 | 103 | 4·3 | ... | ... | S.W. to S.S.E. | 0·2 | ... | | Fine, cloudy. |
| 103 | 103 | ... | 103 | 4·3 | ... | ... | W. to calm | 2 | ... | | Stormy. |
| 103 | 108 | ... | 108 | 4·10 | ... | ... | W. to calm | 1 | ... | | Stormy. |
| ... | ... | ... | ... | ... | ... | ... | ... | ... | ... | | Put to sea for a norther. |
| 50 | 50 | ... | 50 | ... | ... | ... | N. to N.W. | 8·2 | ... | | Heavy sea, rain. |
| ... | ... | ... | ... | ... | ... | ... | N. to calm. | 2 | ... | | Very fine. |
| 200 | 200 | ... | 200 | 9·52 | ... | ... | N. to E. | 2·5 | ... | | Very fine. |
| 200 | 200 | ... | 200 | 9·75 | ... | ... | S.W.—N.—N.W. | 2 | ... | | Very fine. |
| ... | 200 | ... | 200 | 9 09 | ... | ... | N. to N.E. | 2 | ... | | Very fine. |
| 1,342 | 2,179 | 305 | 2,484 | 5·19 | 109·3 | 1 | | | | | |

TH SEA ISLANDS.

| Steam and Sail. | Total Stm. | Sail. | Grand Total Run. | Average Speed per hour. | Tons. | Avge. per Run | Direction. | Force | Hst. | Lst. | Remarks. |
|---|---|---|---|---|---|---|---|---|---|---|---|
| — | ... | — | ... | ... | ... | ... | ... | — | Cumulus | | |
| ... | 114 | ... | 114 | 5 70 | ... | ... | S. to calm. | 2 | Very fine | | Very fine. |
| 52 | 52 | 149 | 201 | 8·37 | ... | ... | S.E. to S. by W. | 5·6 | ... | | Fresh and clear. |
| ... | ... | 200 | 200 | 8·33 | ... | ... | S.S.E. | 5 6 | Cirr. cu. | | Fresh and squally. |
| ... | ... | 185 | 185 | 7·70 | ... | ... | S.S.E. | 6·4 | Cirr. cu. | | Fresh to moderate. |
| ... | ... | 115 | 115 | 4·79 | ... | ... | S.E.—E.—W. by S. | 5·2 | Cumulus | | Light breeze, cloudy. |
| ... | 67 | 18 | 85 | 3·54 | ... | ... | W. by S. | 2 7 | Cumulus | | Light breeze, calm, cloudy. |
| ... | 25 | 148 | 173 | 7·20 | ... | ... | W. by S. to S.E. | 2·7 | Cirro. cu. | | Light to stiff breeze, squally. |
| ... | 15 | 115 | 130 | 5·41 | ... | ... | E. to S.S.E.—W. | 2 | Cumulus | | Light, moderate, fine. |
| 91 | 91 | 27 | 118 | 4·91 | ... | ... | W. by N. to S.S.W. | 8·4 | Cumulus | | Very heavy squalls to fine. |
| ... | ... | 120 | 120 | 5 | ... | ... | S. to S.E. | 4·2 | Cumulus | | Fine, moderate, squally. |
| ... | ... | 101 | 101 | 4·20 | ... | ... | S.E. to E.N.E. | 4·2 | Cumulus | | Fine, moderate, squally. |
| ... | ... | 127 | 127 | 5,3,0 | ... | ... | E. by S. | 2·4 | Cumulus | | Very fine. |
| ... | ... | 162 | 162 | 6·75 | ... | ... | E. by S. to S. E. | 4·2 | Cirro. cu. | | Very fine. |
| ... | — | 167 | 167 | ·7 | ... | ... | E. to E by N. | 5·2 | Cirro. cu. | | Fine, with rain, squally. |
| 143 | 364 | 1,634 | 1,998 | 6·01 | | | | | | | |

*Abstract from Log-book of "Wanderer."*

| 1882. | | | Temperature of Air. Deg. Fahr. (Ht. Lt.) | Temperature of Water. Deg. Fahr. | Position at Noon. Latitude. Deg. | Latitude. Min. | Longitude. Deg. | Longitude. Min. | Course. |
|---|---|---|---|---|---|---|---|---|---|
| Apr. 19 | Leave Beirut ... ... ... ... | 4.40 P.M. | ... | ... | ... | ... | ... | ... | Various |
| | Anchor at Famagousta, Cyprus, April 20 ... ... ... ... | 7.50 A.M. | | | | | | | |
| „ 21 | Leave Famagousta ... ... ... | 5.10 A.M. | ... | ... | ... | ... | ... | ... | Various |
| | Anchor at Larnaka same day ... | 10 25 A.M. | | | | | | | |
| „ 21 | Leave Larnaka .. ... ... ... | 1.30 P.M. | ... | ... | ... | ... | ... | ... | Various |
| | Anchor at Rhodes, April 23 ... | 3.15 P.M. | | | | | | | |
| „ 24 | Leave Rhodes ... ... ... ... | 4 P.M. | ... | ... | ... | ... | ... | ... | Various |
| | Anchor at Smyrna, April 26 ... | 8 A.M. | | | | | | | |
| „ 27 | Leave Smyrna ... ... ... ... | 4 15 P.M. | ... | ... | ... | ... | ... | ... | Various |
| | Anchor at Chenak, April 28 ... | 12 15 P.M. | | | | | | | |
| May. 2 | Leave Chenak ... ... ... ... | 6 30 P.M. | ... | ... | ... | ... | ... | ... | Various |
| | Anch. at Constantinople, May 3 | 9.30 A.M. | | | | | | | |
| „ 6 | Leave Constantinople ... ... | 11 30 A.M. | ... | ... | ... | ... | ... | ... | Various |
| | Anchor at Piræus, May 8 ... | 3 30 P.M. | | | | | | | |
| „ 13 | Leave Salamis Bay ... ... ... | 5.30 A.M. | 59 70 | 63 | 36 | 58 N. | 3 | 32 E. | Various |
| „ 14 | At sea (30·20 30 25) ... ... | ... | 61 74 | 64 | 36 | 21 | 20 | 04 | Various |
| „ 15 | „ (30·00—30 15) ... ... | ... | 59 69 | 63½ | 36 | 19 | 16 | 42 | N. [illegible] W. |
| „ 16 | Arrive at Valletta ... ... ... | 8 A.M. | 62 69 | 65 | ... | ... | ... | ... | Various |
| | *Duration of voyage, 3 days 2 hours.* | | | | | | | | |
| | TOTALS ON VOYAGE ... ... ... | ... | ... | ... | ... | ... | ... | ... | — |
| „ 20 | Leave Malta ... ... ... ... | 4 15 P.M. | 62 75 | 65 | ... | ... | ... | ... | — |
| | Anchor at Syracuse, May 21 ... | 7.45 A.M. | | | | | | | |
| „ 21 | Leave Syracuse ... ... ... | 2 15 P.M. | ... | ... | ... | ... | ... | ... | — |
| | Anchor at Catania, May 21 ... | 6.5 P.M. | | | | | | | |
| „ 23 | Leave Catania ... ... ... ... | 5 A.M. | ... | ... | ... | ... | ... | ... | — |
| | Anchor at Palermo, May 24 ... | 6 A.M. | | | | | | | |
| „ 25 | Leave Palermo ... ... ... | 11 A.M. | ... | ... | ... | ... | ... | ... | — |
| | Anchor at Naples, May 26 ... | 2 25 P.M. | | | | | | | |
| June 1 | Leave Naples ... ... ... ... | 2 P.M. | ... | ... | ... | ... | ... | ... | — |
| | Anchor at Civita Vecchia, June 2 | 2 30 P.M. | | | | | | | |
| „ 9 | Leave Civita Vecchia ... ... | 5 P.M. | 65 75 | 69 | | | | | |
| „ 10 | At sea (30·03—30 15) ... ... ... | Noon | 66 75 | 68 | 40 | 26 N. | 10 | 43 E. | S. [illegible] W. |
| „ 11 | Arrive at Cagliari ... ... ... | 8.45 A.M. | 67 76 | 68 | ... | ... | ... | ... | Various |
| | *Duration of voyage, 1 day 15¾ hours.* | | | | | | | | |
| | TOTALS ON VOYAGE ... ... ... | ... | ... | ... | ... | ... | ... | ... | — |
| „ 12 | Leave Cagliari ... ... ... ... | 8.30 P.M. | 61 79 | 68 | ... | ... | ... | ... | — |
| „ 13 | At sea (30·17—30) ... ... ... | Noon | 66 79 | 68 | 38 | 14 N. | 7 | 19 E. | Various |
| „ 14 | Arrive at Algiers ... ... ... | 8 30 P.M. | 67 73 | 69 | ... | ... | ... | ... | S. 71° W. |
| | *Duration of voyage, 1 day 19 hours.* | | | | | | | | |
| | TOTALS ON VOYAGE ... ... ... | — | — | — | ... | ... | ... | ... | — |
| „ 15 | Leave Algiers ... ... ... ... | 4 P.M. | 65 75 | 68 | ... | ... | ... | ... | — |
| „ 16 | At sea (30 12—30·30) ... ... ... | Noon | 68 71 | 69 | 36 | 47 N. | 0 | 20 E. | Various |
| „ 17 | Arrive at Malaga ... ... ... | 8 P.M. | 69 78 | 68 | 36 | 41 | 3 | 01 W. | W. |
| | *Duration of voyage, 2 days 4 hours.* | | | | | | | | |
| | TOTALS ON VOYAGE ... ... ... | ... | ... | ... | ... | ... | ... | ... | — |
| „ 24 | Leave Malaga ... ... ... ... | 4 15 A.M. | ... | ... | ... | ... | ... | ... | Various |
| | Anchor at Gibraltar same day ... | 2 45 P.M. | | | | | | | |
| July 3 | Leave Gibraltar ... ... ... | 11.45 A.M. | ... | ... | ... | ... | ... | ... | Various |
| | Anchor at Lisbon, July 5 ... | 7.45 A.M. | | | | | | | |
| „ 5 | Leave Lisbon ... ... ... ... | 2.10 P.M. | 70 | 64 | | | | | |
| „ 6 | At sea (30 18—30·30) ... ... | Noon | 64 70 | 65 | 41 | 5 N. | 10 | 8 W. | Various |
| „ 7 | „ (29·80—30·08) ... ... | ... | 63 66 | 63 | 44 | 51 | 9 | 26 | N. 7 E. |
| „ 8 | „ (29 66—29 75) ... ... | ... | 59 65 | 61 | 48 | 50 | 8 | 56 | N. 5 E. |
| „ 9 | Arrive at Queenstown ... ... | 8.20 A.M. | 54 61 | 56 | ... | ... | ... | ... | Various |
| | *Duration of voyage, 3 days 18 hours.* | | | | | | | | |
| | TOTALS ON VOYAGE ... ... ... | ... | ... | ... | ... | ... | ... | ... | — |
| „ 15 | Leave Queenstown ... ... ... | 8 P.M. | ... | ... | ... | ... | ... | ... | Various |
| | Arrive at New Milford, July 16... | 11 A.M. | | | | | | | |
| „ 18 | Leave New Milford ... ... ... | 6 A.M. | ... | 60 | | | | | |
| „ „ | (29·93—30·02) ... ... | Noon | 59 61 | 60 | 50,58 | ... | 5 | 30 | Various |
| „ 19 | „ (30 02—30 26) ... ... | ... | 59 62 | 61½ | 50 | 26 | 2 | 26 | Various |
| „ „ | Arrive at Cowes ... ... ... | 5 P.M. | ... | ... | ... | ... | ... | ... | Various |
| | *Duration of voyage, 1 day 11 hours.* | | | | | | | | |
| | TOTALS ON VOYAGE ... ... ... | ... | ... | ... | ... | ... | ... | ... | — |
| | TOTAL FROM COWES, AUG. 5, 1880, TO JULY 19, 1882 ... ... ... | ... | ... | ... | ... | ... | ... | ... | — |

*...en from Place of Departure to place of Arrival.*

| Distances run under Steam and Sail. | | | | | Average Speed per hour. | Coal used, with average per mile run under steam | | Wind and Weather. | | Remarks. | |
|---|---|---|---|---|---|---|---|---|---|---|---|
| m. | Steam and Sail. | Total Stm. | Sail. | Grand Total Run. | | Tons. | Avge per Run. | Direction. | Force. | | |
| ls. | Mls. | Mls. | Mls. | Mls | Mls. | | Cwt. | | | | |
| 108 | ... | 108 | ... | 108 | 7 2 | | ... | | | | |
| 42 | ... | 42 | ... | 42 | 8·4 | ... | ... | S. by E. | 2 | | |
| 306 | ... | 306 | ... | 306 | 6·12 | ... | ... | N.N.W.—N.W. | 2.4.2 | | |
| 242 | ... | 242 | ... | 242 | 6·05 | ... | ... | Variable | | | |
| 138 | ... | 138 | ... | 138 | 6·9 | | | | | | |
| 132 | ... | 132 | ... | 132 | 8·8 | | | | | | |
| 381 | ... | 381 | ... | 381 | 7·32 | | | | | | |
| ... | 59 | 59 | ... | 59 | 9·07 | ... | ... | N N.W. | 4 | Cirro. cu. | |
| 165 | ... | 165 | ... | 165 | 6·87 | ... | ... | N.—c—N.W. by N. | 4.0.2 | Cirro. str. cu. | |
| ... | 180 | 180 | ... | 180 | 7·5 | ... | ... | W.—S.W. by W.—S.W. | 4.2.4 | Cirro. cumulus | |
| 108 | ... | 108 | ... | 108 | 5·4 | 85·3 | ·804 | S.W.—c. | 2 0 | Nim. overcast | |
| 278 | 239 | 512 | ... | 512 | 6·92 | ... | ... | ... | ... | ... | Av. per diem, 166·06 mls. |
| 84 | ... | 84 | ... | 84 | 5·42 | ... | ... | N.E. by N.—W. | 4.2 | Overcast, cir. cu. | |
| 32 | ... | 32 | ... | 32 | 8 | | | | | | |
| 124 | ... | 124 | ... | 124 | 5 | ... | ... | ... | ... | ... | Looked in at Milazzo, to see tunny-fishing. |
| 170 | ... | 170 | ... | 170 | 6·3 | | | | | | |
| 146 | ... | 146 | ... | 146 | 6·08 | | | | | | |
| 110 | ... | 110 | ... | 110 | 5·8 | ... | ... | W.—W S.W.—S.W. by W. | 4.6 5 | Cir. cu. cir. str. | |
| 116 | ... | 116 | ... | 116 | 5·8 | ... | ... | S.S W.—W—N.N.W. | 2.6 | Cirrus, cir. str. | |
| 226 | ... | 226 | ... | 226 | 5·8 | ... | ... | ... | ... | ... | Av. per. diem, 139·2 mls. |
| ... | ... | ... | ... | ... | ... | ... | ... | ... | ... | Cirro. stratus | |
| ... | 102 | 102 | ... | 102 | 6·57 | ... | ... | N. by W—c—S.E by E. | 2.0 2 | Cirro. cu. | |
| ... | 238 | 238 | ... | 238 | 8·47 | ... | ... | E.N.E.—N.E.—E. | 5 2 | Cirro. cu. | |
| ... | 335 | 335 | ... | 335 | 7·3 | ... | ... | ... | ... | ... | Av. per diem, 187·2 mls |
| ... | ... | ... | ... | ... | ... | ... | ... | ... | ... | Cirro. cu. | |
| 131 | ... | 131 | ... | 131 | 6·55 | ... | ... | W—W S W. | 2 4.2 | Overcast | |
| 229 | ... | 229 | ... | 229 | 7·15 | ... | ... | W—N W—E.S.E. | 2.4 | Cirro stratus | |
| 360 | ... | 360 | ... | 360 | 6·92 | ... | ... | ... | ... | ... | Av. per diem, 166·08 mls. |
| 65 | ... | 65 | ... | 65 | 6·19 | 54·8 | ·706 | | | | |
| 306 | ... | 306 | ... | 306 | 6·95 | ... | ... | | | | |
| ... | 175 | 175 | ... | 175 | 8·33 | ... | ... | N.W—W. by S. | 2.4 | Cir. cu. overcast | |
| ... | 228 | 228 | ... | 228 | 9·5 | ... | ... | N.W—W—W.N W. | 4 5 | Overcast, nim | |
| ... | 240 | 240 | ... | 240 | 10 | ... | ... | W.N.W—N.N.W.—N.W. | 5 | Nim. overcast | |
| 167 | ... | 167 | ... | 167 | 8·35 | ... | ... | N.W.—W N.W. | 5.4 | Nim. overcast | |
| 167 | 643 | 810 | ... | 810 | 9 | ... | ... | ... | ... | ... | Av. per diem, 216 mls. |
| ... | 173 | 173 | ... | 173 | 11·53 | ... | ... | | | | |
| 59 | ... | 59 | ... | 59 | 9·83 | ... | ... | W.S.W.—S.W. | 5.6 | Cumulus | |
| ... | 190 | 190 | ... | 190 | 7·92 | ... | ... | S.W.—S.S.W. | 6 | Cumulus | |
| 43 | ... | 43 | ... | 43 | 8·6 | 65·4 | ·83 | | | | |
| 92 | 190 | 292 | ... | 292 | 8·34 | ... | ... | ... | ... | ... | Av. per diem, 200·16 mls. |
| 13,186 | 17,829 | 31,015 | 18,875 | 48,490 | 6·68 | ... | ... | ... | ... | ... | Av. per diem, 160·32 mls. |

| Date and Name of Places of Departure and Arrival. | | | | Time of Leaving and Arriving and Duration of Voyage. | | | | Dista- |
|---|---|---|---|---|---|---|---|---|
| Date. 1880. | Place of Departure. | Time. | Place of Arrival. | Date and Time of Arrival. Date. | Time. | Duration of Voyage. Ds. | Hs. | Steam Miles. |
| Aug. 5 | Cowes... | 7.15 P.M. | Vigo | Aug. 9 | 6.30 A.M. | 3 | 12 | [illegible] |
| ,, 11 | Vigo | 11.35 A.M. | Lisbon | ,, 12 | 11.35 A.M. | 1 | 0 | [illegible] |
| ,, 21 | Lisbon | 12 A.M. | Funchal | ,, 25 | 12 A.M. | 4 | 0 | [illegible] |
| ,, 27 | Funchal | 11 A.M. | Gaboon | Sep. 17 | 2 P.M. | 21 | 3 | [illegible] |
| Sep. 26 | Gaboon | 6.30 A.M. | St. Helena | Oct. 4 | 6 30 A.M. | 8 | 0 | [illegible] |
| Oct. 7 | St. Helena | 10.35 A.M. | Bahia | ,, 20 | 4 P.M. | 13 | 6 | [illegible] |
| ,, 26 | Bahia | 5 P.M. | Rio Janeiro | ,, 31 | 8 P.M. | 4 | 12 | [illegible] |
| Nov. 13 | Rio Janeiro | 6 P.M. | Monte Video | Nov. 19 | 8 P.M. | 6 | 6 | 343 |
| ,, 22 | Monte Video | 3 P.M. | Buenos Ayres | ,, 23 | 7 A.M. | 0 | 15 | [illegible] |
| ,, 30 | Buenos Ayres | 5 P.M. | Monte Video | Dec. 1 | 7 A.M. | 0 | 14 | [illegible] |
| Dec. 4 | Monte Video | 5 P.M. | Condor Cliff, Cape Virgins | ,, 16 | 6.30 A.M. | 8 | 2 | [illegible] |
| ,, 19 | Condor Cliff | 2.30 A.M. | Sombrero Island abeam | 1881. Jan. 7 | 8 P.M. | 5 | 4 | [illegible] |
| 1881 Jan. 8 | Sombrero Isl., Jan. 7 | 8 P.M. | Coquimbo | ,, 13 | 3 P.M. | 5 | 14 | [illegible] |
| | Totals and Total Averages from Cowes to First Arrival at Coquimbo, Chili | | | | | 81 | 10 | [illegible] |

*Deduct Coal used during Voyage.—By Banked Fires, 26 Tons; by Galley, 12 Tons; Distilling, 3½ Tons:* ...

Time under Way, propelled by Steam with or without the aid of Sails ... ... — — — —

Time under Way, propelled by Sails ... ... ... ... ... ... ... ... ... — — —

Time in Harbours ... ... ... ... ... ... ... ... — — ... — — — —

Total. On the Voyage between leaving Cowes on Aug. 5, 1880, and arrival in Coquimbo, Jan. 13, 1881 —

*NOTE.—During our stay in Harbours our Fires have been banked at Funchal, Madeira, for 2 days,* ...

WORK DONE DURING OUR STAY IN CHILI, WHERE WE ARRIVED ON J...

| Date | Place of Departure | Time | Place of Arrival | Date | Time | Ds. | Hs. | Steam Miles |
|---|---|---|---|---|---|---|---|---|
| Feb. 15 | Coquimbo | Noon | Coquimbo | June 9 | 2.30 P.M. | 19 | 22 | [illegible] |
| June 16 | Coquimbo | 4 P.M. | Anchor Bon Repos B., Fatou Hiva, Marquesas | July 14 | 2.55 P.M. | 27 | 23 | [illegible] |
| July 16 | Fatou-Hiva | 6 A.M. | Obelisk Island, Roa-Poua | ,, 19 | 5.45 P.M. | 1 | 8½ | 56 |
| ,, 19 | Obelisk Island | 5.45 P.M. | Tahiti | ,, 24 | 4.30 P.M. | 5 | 22¾ | [illegible] |
| Aug. 6 | Tahiti | 11 A.M. | Bora-bora | Aug. 12 | 12 30 P.M. | 1 | 1½ | 52 |
| ,, 15 | Bora-bora | 7 A.M. | Off Rarotonga | ,, 18 | Noon | 3 | 5 | — |
| ,, 18 | Off Rarotonga | Noon | Tonga-Tabou | ,, 24 | 1.45 P.M. | 6 | 1¾ | — |
| ,, 28 | Tonga-tabu | 6.30 A.M. | Levuka, Ovalua, Fiji | ,, 31 | 4.30 P.M. | 2 | 10 | — |
| Sep. 11 | Levuka | 7 A.M. | Suva, Viti Levu | Sept. 11 | 3 P.M. | 0 | 8 | — |
| ,, 15 | Suva | 9.56 A.M. | Honolulu, Oahu | Oct. 7 | 10.30 A.M. | 22 | 0 | 151 |
| Oct. 20 | Honolulu | 11 A.M. | Wailuku, Maui | Nov. 15 | 10 A.M. | 3 | 14 | [illegible] |
| Nov. 16 | Wailuku Bay, Maui | 3 P.M. | Yokohama | Dec. 11 | Noon | 23 | 20 | — |
| Dec. 21 | Yokohama | 8 A.M. | Kobe | ,, 23 | 8 A.M. | 2 | 0 | 35 |
| ,, 29 | Kobe | 11 A.M. | Simonoseki | 1882. Jan. 2 | 11 P.M. | 1 | 12 | 296 |
| 1882 Jan. 3 | Simonoseki | 2.15 P.M. | Fusan, Korea | ,, 4 | 5.15 A.M. | 0 | 15 | 120 |
| ,, 4 | Fusan | 5 P.M. | Nagasaki | ,, 6 | 2 15 P.M. | 1 | 3 | 45 |
| ,, 9 | Nagasaki | 8 A.M. | Hong-Kong | ,, 15 | 11 A.M. | 6 | 3 | 133 |
| Feb. 7 | Hong-Kong | Noon | Singapore | Feb. 16 | 5.45 P.M. | 9 | 5 | 154 |
| ,, 21 | Singapore | 10.5 A.M. | Johore | ,, 21 | 3.10 P.M. | 0 | 5 | 33 |
| ,, 22 | Johore | 10.30 A.M. | Singapore | ,, 22 | 4.20 P.M. | 0 | 5½ | 33 |
| ,, 22 | Singapore | 5 30 P.M. | Malacca | ,, 23 | 12.15 P.M. | 0 | 18¾ | 125 |
| ,, 23 | Malacca | 5 55 P.M. | Colombo | Mar. 5 | 8.20 A.M. | 9 | 16 | [illegible] |
| Mar. 9 | Colombo | 3.40 P.M. | Suez | ,, 28 | Noon | 18 | 21 | 1,021 |
| ,, 29 | Suez | Noon | Port Said | April 4 | 6 20 P.M. | 1 | 4 | [illegible] |
| April 6 | Port Said | 4.25 P.M. | Gibraltar | June 24 | 2 45 P.M. | 22 | 16½ | 3,066 |
| July 3 | Gibraltar | 11.45 A.M. | Lisbon | July 5 | 7.45 A.M. | 1 | 20 | 396 |
| ,, 5 | Lisbon | 2.10 P.M. | Queenstown | ,, 9 | 8.20 A.M. | 3 | 18 | 167 |
| ,, 15 | Queenstown | 8 P.M. | New Milford | ,, 16 | 11 A.M. | 0 | 15 | — |
| ,, 18 | New Milford | 6 A.M. | Cowes | ,, 19 | 5 P.M. | 1 | 11 | 59 |
| | | | | | | 280 | 0 | 13,196 |

Duration of Voyage *1 year 348 days*; out of which *1 year and 68 days* in Harbour, and *280 days* at ...

| ler Steam and Sail. | | | Average Speed. | | Coal consumed. | | | Coal received on board. Tons. | Direction and Strength of Wind. | | Days under Steam. |
|---|---|---|---|---|---|---|---|---|---|---|---|
| otal eam. iles. | Sail. Miles. | Grand Total Miles. | Per Diem Miles. | Per Hour. Miles. | During Passage. Tons. | Per Diem. Tons. | Average per Mle. Cwts. & Decls. | | Direction. | Strength. | Ds. Hs. |
| 721 | ... | 721 | 206 | 8·58 | 34 | 9 75 | 0·943 | 95 | ... | ... | 3 12 |
| 257 | ... | 257 | 257 | 10·70 | 16·25 | 16·25 | 1·264 | ... | ... | ... | 1 0 |
| 257 | 180 | 537 | 134½ | 5·59 | 11·50 | 4·60 | 0·684 | 55 | ... | ... | 2 12 |
| 527 | 2.122 | 3,649 | 173 | 7 20 | 52·25 | 6·14 | 0·681 | ... | ... | ... | 8 12 |
| 906 | 476 | 1,382 | 172½ | 7·17 | 40·50 | 7·36 | 0·894 | 44 | ... | ... | 5 12 |
| 656 | 1,263 | 1,919 | 144½ | 6·03 | 28·25 | 7·53 | 0·861 | ... | ... | ... | 3 18 |
| 753 | ... | 753 | 167 | 6·78 | 23·50 | 5·22 | 0·624 | 84 | ... | ... | 4 12 |
| 045 | ... | 1,045 | 174 | 7·15 | 41 | 6·83 | 0·669 | ... | ... | ... | 6 2 |
| 115 | ... | 115 | 171·60 | 7·66 | 7·75 | 10·33 | 1·343 | ... | ... | ... | 0 15 |
| 115 | ... | 115 | 197 | 8 21 | 7·25 | 9·66 | 1·260 | ... | ... | ... | 0 14 |
| ,394 | ... | 1,394 | 174·25 | 7·18 | 53·25 | 6 45 | 0·764 | 86 | ... | ... | 8 2 |
| 917 | ... | 917 | 177 12 | 7·88 } | 113·50 | 10·31 | 1·127 | 70 | Taken at Pa. Arenas. | ... | { 5 4 |
| 097 | ... | 1,097 | 196 3 | 8·18 } | | | | | | | { 5 14 |
| ,860 | 4,041 | 13,901 | 170·56 | 7·11 | 429·) | 7·73 | 0·870 | 434 | ... | ... | 55 9 |
| *utter, 2¼ Tons = to* | | | ... | ... | 385·6 | 6·93 | 0·780 | — | — | — | — |

55 Days 9 Hours, with an Average Speed per Hour of 7·41 Knots, upon a run of 9,860 Geographical Miles.
26 Days 1 Hour, with an Average Speed per Hour of 6·46 Knots, upon a run of 4,041 Geographical Miles.
79 Days 14 Hours. Average Steam and Sail, 7·11 Knots, upon a run of 13,901 Geographical Miles.

161

*ur arrival at Monte Video, Nov. 19, to our leaving Port Grappler, Jan.* 7, 1881, 36 *days. Total* 38 *days.*

THURSDAY) AND LEFT AGAIN ON JUNE 16, MAKING A STAY THERE OF 22 WEEKS.

| Total Steam Miles | Sail Miles | Grand Total Miles | Per Diem Miles | Per Hour Miles | During Passage Tons | Per Diem Tons | Average per Mle. | Coal received | Direction | Strength | Ds. Hs. |
|---|---|---|---|---|---|---|---|---|---|---|---|
| ,179 | 305 | 2,484 | 124·56 | 5·19 | 109·3 | 6·16 | 1 | 207·10 | ... | ... | 17 17 |
| 377 | 3,633 | 4,010 | 143·38 | 5·976 | ... | ... | ... | ... | ... | ... | 2 23 |
| 180 | ... | 180 | 133·92 | 5·58 | Having touched at Vaitahu B., Jacuata and Jaie-O-Hae, Nouka-Hiva, and coasted Roa-Poua, to Obelisk Id. | | | | | | 1 8½ |
| | | | | | 46·8 | 7·66 | 1·13 | 40 | | | |
| 287 | 437 | 724 | 146·16 | 6·09 | 12 hours under low steam waiting for daylight off Ahii Atoll, and 5 hours' delay in approaching and landing | | | | | | 1 23½ |
| 174 | ... | 174 | 165·36 | 6·89 | Being visits to Taloo Har., Moorea and Utuvoa Har., Raiatea, at each of which places we made some stay. | | | | | | 1 1¼ |
| ... | 545 | 545 | 169·68 | 7·07 | Thick and stormy weather; wind blowing on small entrance to harbour, compelled us to abandon visit to this place. | | | | | | |
| 900 | ... | 900 | 146·4 | 6·10 | ... | ... | ... | ... | ... | ... | 6 1½ |
| 417 | ... | 417 | 166·8 | 7·18 | ... | ... | ... | ... | ... | ... | 2 10 |
| | | | | | 55·1 | 5·6 | 7·11 | 59·10 | ... | ... | |
| 56 | ... | 56 | 196 | 8 | 1 hour break down of engines, therefore averg. speed is taken on 7 hours | | | | | | 0 8 |
| 3,068 | 99 | 3,167 | 143·95 | 5·99 | 122·12 | 4·45 | ·799 | 127·2 | ... | ... | 22 0 |
| 568 | ... | 568 | 158 65 | 6·6 | Calling twice at Honolulu, also at Kealakeakua B., Hilo, and Kailuku B. | | | | | | 3 14 |
| 1,325 | 2,792 | 4,117 | 171·54 | 7·18 | 69 | 6·42 | ·775 | 50 | ... | ... | 7 21 |
| 374 | ... | 374 | 187 | 7·79 | ... | ... | ... | ... | ... | ... | 2 0 |
| 286 | ... | 286 | 190 | 7·91 | Calling at Sakate, Mitaria, Miyashima and Kaminoseki in Inland Sea. | | | | | | 1 12 |
| 120 | ... | 120 | ... | 8 | ... | ... | ... | ... | ... | ... | 0 15 |
| 180 | ... | 180 | 159·84 | 6·66 | Calling at Hiradorio-seto, Spec. Str. | | | | ... | ... | 1 3 |
| 1,045 | ... | 1,045 | 170·4 | 7·1 | 82 18 | 7·29 | ·826 | 96·14 | ... | ... | 6 3 |
| 324 | 1,137 | 1,461 | 158·64 | 6·61 | 18·11 | 7·24 | 1·12 | 20 | ... | ... | 2 5 |
| 33 | ... | 33 | ... | 6·6 | ... | ... | ... | ... | ... | ... | 0 5 |
| 33 | ... | 33 | ... | 6 | ... | ... | ... | ... | ... | ... | 0 5½ |
| 125 | ... | 125 | ... | 6·58 | ... | ... | ... | ... | ... | ... | 0 18½ |
| 502 | 956 | 1,458 | 150·93 | 6·28 | 24·8 | 6·2 | ·704 | 32 14 | ... | ... | 2 17 |
| 3,274 | 180 | 3,404 | 186·72 | 7·78 | ... | ... | ... | ... | ... | ... | 17 21 |
| | | | | | 103·7 | 5·45 | ·614 | 89 | ... | ... | 1 4 |
| 87 | ... | 87 | 86·88 | 3·62 | Calling at Little Bitter Lake and Ismailia. | | | | ... | ... | |
| | | | | | 189·11 | 6·31 | ·76 | 129 | | | |
| 3,660 | ... | 3,660 | 161·52 | 6·78 | Calling at Jaffa, Beirut, Famagousta, Larnaka, Rhodes, Smyrna, Chenak, Constantinople, Piræus, Salamis B., Valletta, Syracuse, Catania, Palermo, Naples, Civita Vecchia, Cagliari, Algiers, and Malaga. | | | | | | 22 16½ |
| 306 | ... | 306 | 166·8 | 6·95 | ... | ... | ... | ... | ... | ... | 1 20 |
| 810 | ... | 810 | 216 | 9 | ... | ... | ... | ... | ... | ... | 3 18 |
| 173 | ... | 173 | ... | 11 53 | ... | ... | ... | ... | ... | ... | 0 15 |
| 292 | ... | 292 | 200·16 | 8·34 | 65·4 | 9 3 | ·83 | ... | ... | ... | 1 11 |
| 31,015 | 13,875 | 44,890 | 160·82 | 6 68 | 1,265·10 | 6·7 | ·816 | 1.265·10 | ... | ... | 189 14½ |

during which we traversed *44,890 Geographical Miles*, with the Consumption of 1,265·10 Tons of Coal.

LONDON:
R. CLAY, SONS, AND TAYLOR.
BREAD STREET HILL.

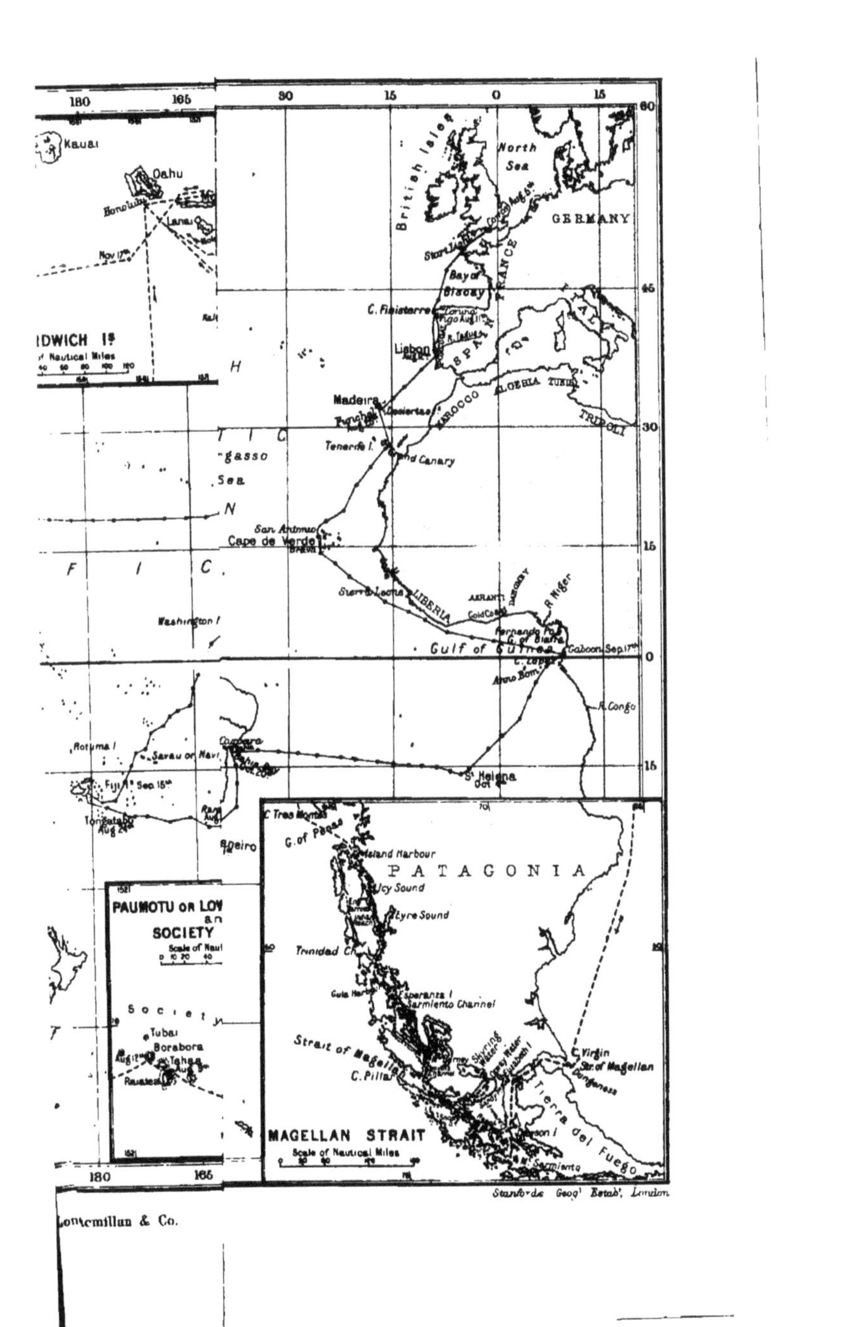

North Sea
GERMANY
British Isles
FRANCE
Bay of Biscay
C. Finisterre
Lisbon
SPAIN
ITALY
ALGERIA
TUNIS
TRIPOLI
MAROCCO
Madeira
Funchal
Desiertas
Teneriffe I.
Grand Canary
Sargasso Sea
San Antonio
Cape de Verde
Brava
Sierra Leone
LIBERIA
ASHANTI
DAHOMEY
R. Niger
Gold Coast
Fernando Po
G. of Biafra
Gulf of Guinea
Gaboon Sep. 17th
C. Lopez
Anno Bom
R. Congo
St. Helena
Washington I.
Kauai
Oahu
Honolulu
Lanai
Nov. 17th
SANDWICH Is.
Nautical Miles
Rotuma I.
Sarau or Navi.
Fiji Is. Sep. 15th
Tongatabu Aug. 24th
PAUMOTU OR LOW
SOCIETY
Scale of Nautical
Society
Tubai
Borabora
Tahaa
Raiatea
C. Tres Montes
G. of Peñas
Island Harbour
PATAGONIA
Icy Sound
Eyre Sound
Trinidad Ch.
Esperanza I.
Sarmiento Channel
Strait of Magellan
C. Pillar
C. Virgin
Str. of Magellan
Dungeness
Tierra del Fuego
MAGELLAN STRAIT
Scale of Nautical Miles
Stanford's Geog'l Estab't, London
Macmillan & Co.

www.ingramcontent.com/pod-product-compliance
Ingram Content Group UK Ltd.
Pitfield, Milton Keynes, MK11 3LW, UK
UKHW021448020425
5276UKWH00041B/841

9 781016 403771